Agricultural Finance and Management

Agricultural Finance and Management

Vikram Sharma

RANDOM

RANDOM PUBLICATIONS

NEW DELHI (INDIA)

Agricultural Finance and Management

ISBN 978-93-5111-361-4

Published in 2014 in India by

RANDOM PUBLICATIONS

4376-A/4B, Gali Murari Lal, Ansari Road
New Delhi-110 002
Phone : +91-11-43580356, +91-11-23289044
e-mail: randomexports@gmail.com, sales@randompublications.com, info@randompublications.com

Reprinted 2025

Type Setting by : Keystoneprintads, Delhi-110051
Digitally Printed at : Replika Press Pvt. Ltd.

Preface

The use of sound financial management practices in agriculture is quickly becoming necessary for survival and are a required for outstanding economic performance. We developed a research program to determine how widely farmers had adopted various financial management practices and whether these practices had any impact on farm profitability. Financing for agriculture is viable and sustainable if supported by sound risk assessment and risk management at all levels, including at the farm, the financial institution (commercial risk), and the agricultural value chain.

Cooperative financial institutions and mutuals are often well positioned and effective in providing financial services for agriculture, and other financial institutions can draw valuable lessons from their experiences. Innovations in agricultural finance could build on and improve existing products and processes to deliver new and more efficient services. Investment costs of innovation may deter good initiatives. Cost-sharing, through the development of partnerships for example with other service providers, is necessary both to initiate and to scale up innovation. Finance in agriculture is as important as development of technologies.

Technical inputs can be purchased and used by farmer only if he has money (funds). But his own money is always inadequate and he needs outside finance or credit.

The Indian commercial banks are mandated by the Reserve Bank of India (RBI), to undertake directed lending for agriculture and rural development. Commercial banks are required to achieve priority sector lending of 40% of net bank credit of which 18% should be for agriculture.

Agricultural marketing involves in its simplest form the buying and selling of agricultural produce. In olden days when the village economy was more or less self-sufficient the marketing of agricultural produce presented no difficulty as the farmer sold his produce to the consumer on a cash or barter basis.

This book is supposed to serve as a tool for the students, researchers and scholars in the field of Agriculture studies.

I thank all members of my team who have helped in the preparation of the book. My special thanks go to "Random Publications" who have published the book.

– Vikram Sharma

Contents

1

Agriculture Finance

Finance in agriculture is as important as development of technologies. Technical inputs can be purchased and used by farmer only if he has money (funds). But his own money is always inadequate and he needs outside finance or credit. Professional money lenders were the only source of credit to agriculture till 1935. They use to charge unduly high rates of interest and follow serious practices while giving loans and recovering them. As a result, farmers were heavily burdened with debts and many of them perpetuated debts. There were widespread discontents among farmers against these practices and there were instances of riots also. With the passing of Reserve Bank of India Act 1934, District Central Co-operative Banks Act and Land Development Banks Act, agricultural credit received impetons and there were improvements in agricultural credit.

A powerful alternative agency came into being. Large-scale credit became available with reasonable rates of interest at easy terms, both in terms of granting loans and recovery of them. Although the co-operative banks started financing agriculture with their establishments in 1930's real impetons was received only after Independence when suitable legislation were passed and policies were formulated. Thereafter, bank credit to agriculture made phenomenal progress by opening branches in rural areas and attracting deposits. Till 14 major commercial banks were nationalized in 1969, co-operative banks were the main institutional agencies providing finance to agriculture. After nationalization, it was made mandatory for these banks to provide finance to agriculture as a priority sector. These banks undertook special programmes of branch expansion and created a network of banking services throughout the country and started financing agriculture on large scale. Thus agriculture credit acquired multi-agency dimension. Development and adoption of new technologies and availability of finance go hand in hand. In bringing "Green Revolution", "White Revolution" and now "Yellow Revolution" finance has played a crucial role. Now the agriculture credit, through multi agency approach has come to stay. The procedures and amount of loans for various purposes have been standardized. Among the various purposes "Crop loans" (Short-term loan) has the major share. In addition, farmers get loans for purchase of electric motor with pump, tractor and other machinery, digging wells or boring wells, installation of pipe lines, drip irrigation, planting fruit orchards, purchase of dairy animals and feeds/fodder for them, poultry, sheep/goat keeping and for many other allied enterprises.

SCHEMES FOR AGRICULTURE FINANCE

SBT Kisan Gold Card Scheme (General purpose Agriculture Term Loan)
Eligibility:

- Farmers having good track record of repayment for the last two years.
- Farmers who have closed their loan account without default and not our current borrowers.
- Farmers who have defaulted in repayment but closed the Loan within the stipulated repayment period.
- Farmers who are maintaining deposits with the Bank.
- Good borrowers of other banks provided they liquidate their dues with other banks.
- Good farmers who have not availed loans from any bank.
- *Purpose*: The borrower is at liberty to utilize 50 per cent of the amount for any purpose, including consumption purpose and purchase of land.
- *Amount of Loan*: The amount of loan is limited to five times the annual farm income including income from allied activities or 50 per cent of the value of the land offered as collateral security, whichever is less, subject to a maximum of ₹10 lakh.
- *Rate of Interest*: Interest rate ranges from 1 per cent below PLR.
- *Security*: Hypothecation of crops and assets, if any, created out of bank finance and existing movable assets such as milch animals, pump sets etc. The loan will be secured by equitable mortgage of properties worth double the loan amount, or term deposit receipts, LIC policies of adequate surrender value, NSCs completed lock in period or more etc.
- *Disbursement*: Cash disbursals are allowed to the full extent of the credit limit.
- *Repayment*: The repayment period shall be 10 years. The due date of the installment shall be fixed in such away to coincide with the date of generation of income.

HOMESTEAD FARMING

- *Purpose*: A scheme for financing farmers practicing mixed cropping/intercropping along with allied activities to enable them to undertake cultivation of various crops in a more integrated way. The scheme provides the farmers with sufficient working capital required for their homestead farming (Mixed cropping along with allied activities) by fixing scale of finance based on land holding to meet the cost of entire farming activities.

- *Amount of Loan*: The farmers who own cultivated land below one acre be given the scale of finance on pro rata basis at the rate of ₹ 40000/- and farmers who own more than one acre of land be given at the rate of ₹ 37500/- per acre and part thereof.
- *Rate of Interest*: Interest rate ranges from 2.50 per cent below to 1.50 per cent above BPLR for various limits.
- *Repayment*: The facility will be sanctioned as an Agriculture Cash Credit limit (In case of Kisan Credit Card running cash credit).

LOAN FOR ESTATE PURCHASE

- *Eligibility*: The estate should be either in yielding stage with the crops in its prime yield age or capable of being developed in to a viable unit. The yield/ net income of the estate should be sufficient to liquidate the proposed loan and interest accrued with in a period of 7 to 10 years. The proposed estate should be free from encumbrance and entire property should be offered as security to the loan.
- *Purpose*: To encourage those who prefer to settle down in agriculture and are in the look out of good/viable estates for purchase and also to improve production in agriculture.
- *Amount of Loan*: The quantum of loan that will be considered for sanction will be 75 per cent of the registered value or 50 per cent of the market value whichever is low. In exceptional cases 80 per cent of the registered value or 50 per cent of the market share whichever is low is also considered. The loan for the development of the estate like land development including working capital can also be sanctioned.
- *Rate of Interest*: Interest rate same as BPLR.
- *Repayment*: Repayment of loan will be in quarterly/half yearly/yearly installments depending on the harvest of the crops and the loan shall be repaid within a maximum period of 7 to 10 years.

KISAN CREDIT CARD SCHEME

- *Eligibility*: All agriculturists who are in need of short term production requirements. ATM facility and Personal Accident Insurance Scheme for life up to ₹ 50000 and permanent disability cover up to ₹ 25000 is available on request.
- *Purpose*: To provide hassle free short-term credit to farmers on the basis of their land holdings for purchase of inputs and draw cash to meet their production needs. *i.e.* Cultivation expenses including allied activities with a consumption component.
- *Amount of Loan*: To be fixed on the basis of operational holdings and scale of finance with consumption component 15 per cent

(maximum ₹ 10000/-) of production credit. The scale of finance to farmers who own cultivated land below one acre will be at the rate of ₹ 40000/- (on pro rata basis) and farmers who own more than one acre with intensive farming of land be given at the rate of ₹ 37500/- per acre and part thereof.

- *Rate of Interest*: Interest rate ranges from 2.50 per cent below to 1.50 per cent above BPLR for various limits.
- *Repayment*: Running Cash Credit account for 36 months subject to annual review and total annual credit should exceed annual debit.

SCHEME FOR FINANCING FARMERS FOR PURCHASE OF LAND FOR AGRICULTURAL PURPOSES

- *Eligibility:* Small and Marginal farmers - land maximum up to 5 acres of non-irrigated land or 2.5 acres of irrigated land including the land purchased under the scheme. Tenant, sharecropper and landless agricultural labourers with a good record of prompt repayment of our loans for the last 2 years are also eligible.
- *Purpose*: To finance small and marginal farmers, share croppers, tenant cultivators for purchasing land to expand activities and to make existing small and marginal units economically viable to bring fallow lands and waste lands under cultivation to step up agricultural production as well as productivity also to finance share croppers/ tenant farmers to enable them to diversify farming activities to allied areas to increase their income.
- *Amount of Loan*: Maximum loan under the scheme towards land cost shall not exceed ₹ 5 lakh. Cost of development/economic activity shall be financed under the bank's other financing schemes.
- *Rate of Interest*: Interest rate ranges from 1.75 per cent below to 2.00 per cent above BPLR for various limits.
- *Repayment*: Repayment of the loan will be 7 to 12 years in half yearly/ yearly installments with maximum of 24 months moratorium period. Gestation period/repayment due dates etc. will be fixed according to income generation from the activity.

PRODUCE MARKETING LOAN (ADVANCE AGAINST WAREHOUSE RECEIPT)

- *Eligibility*:
  a. Farmers/ traders depositing farm produce in the warehouses of the central/ state warehousing corporations.
  b. Scheme will be operative in Karnataka, Andhra Pradesh, Tamilnadu and Kerala.
- *Purpose*:
  a. To protect the farmers from the compulsion to sell their produce immediately after harvest of produce despite an adverse market.

b. To finance farmers and traders against warehouse receipt.

- *Amount of Loan*: 70 per cent of the value of the warehouse receipt, valued at the market value or 70 per cent of the market price advised by Agri. Dept, HO whichever is less.
- *Rate of Interest*
  Farmers:
  - Up to `3 lakh - 3.50 per cent below PLR 9.50 per cent
  - Above `3 lakh - 2.50 per cent below PLR 10.50 per cent

  Traders:
  - 2.50 per cent below PLR 10.50 per cent (Irrespective of the limit)
- *Repayment:* On demand/6 months which can be extended up to 12 months subject to satisfactory shelf life/market condition.

AGRI-LOAN TO NON-RESIDENT INDIANS

- *Eligibility*: Agricultural advances are available to the resident family members (means spouse, father, mother, brother, sister etc.) of Non-Resident Indians for land-based activities in respect of the land held by them in India subject to:
  a. The loan should be need based and the total land holding of the Non-Resident Indian, in individual name or jointly with others, should not exceed 5 ha.
  b. The loan amount shall not be used for acquiring any additional land.
- *Purpose:* To finance farmers only for land-based activities and to carryon agricultural activities on the existing land.
- *Amount of Loan*: The maximum amount of the loan will be need based.
- *Rate of Interest*: Interest rate ranges from 2.50 per cent below to 1.50 per cent above BPLR for various short-term limits and from 1.75 per cent below to 2.00 per cent above BPLR for various long-term limits.
- *Repayment*: The loan can be repaid out of the income generated from the agricultural activities or remittances from abroad or by debit to their NRE/NRO/FCNR accounts.

MINOR IRRIGATION

Projects with cumulative command area of less than 2000 ha are called minor irrigation projects:

- *Eligibility*: The beneficiary should have a minimum of 50 cents of land to be brought under irrigation to ensure viability and repayment of loan.
- *Purpose*: Scheme for developing irrigation potential, Minor Irrigation, Installation of Pump set Drip Irrigation etc.

- *Amount of Loan*: As per the project submitted.
- *Rate of Interest*: Interest rate ranges from 1.75 per cent below to 2.00 per cent above BPLR for various limits.
- *Repayment*: The loan shall be repaid within a period of 9 years, in yearly installments.

FARM MECHANISATION

Loan for Farm Mechanisation, Purchase of tractors, Power Tillers, etc.

- *Eligibility*:
  a. Tractors with engine capacity up to 35 HP – The applicant should own/ cultivate six acres of perennially irrigated land.
  b. Tractors with engine capacity above 35 HP – The applicant should own/ cultivate eight acres of perennially irrigated land.
  c. Power Tillers – the applicant should own/ cultivate four acres of perennially irrigated land.
- *Purpose*: To purchase tractor/power tillers for agricultural activities.
- *Amount of Loan*: Amount of advance will be the investment cost of tractor/ power tiller and implements less margin @15 per cent.
- *Rate of Interest*: Interest rate ranges from 1.75 per cent below to 2.00 per cent above BPLR for various limits.
- *Repayment*: The period of repayment shall be 9 years for tractors and 7 years for power tillers.

SCHEME FOR CULTIVATION OF MEDICINAL PLANTS

- *Eligibility*: All agriculturists are eligible.
- *Purpose*: Scheme for financing cultivation of 22 medicinal plants cultivated extensively and also in great demand in the local as well as foreign market.
- *Amount of Loan*: Depending on the area of cultivation/ project cost
- *Rate of Interest*: Interest rate ranges from 1.75 per cent below to 2.00 per cent above BPLR for various limits.
- *Repayment*: Repayment should coincide with harvesting and marketing or at the time generation of income from the scheme.

SCHEME FOR CULTIVATION OF VANILLA

- *Eligibility*: All agriculturists are eligible.
- *Purpose*: Scheme for financing cultivation of Vanilla, a cash crop, gaining ground in the State of Kerala.
- *Amount of Loan*: Amount of finance will be ₹250000 per hectare for pure crops and ₹210000/- per hectare for intercrop.
- *Rate of Interest*: Normal rate of interest as applicable to ATL.
- *Repayment*: The loan shall be repaid within a period of 7 years, in yearly installments. Farmer's eligible for two years gestation period

and interest is repayable on the 3rd and 4th year and the principal from the 5th to 7th year.

SBT RAINWATER HARVESTING SCHEME

- *Eligibility*: Farmers having land holding of 0.50 acre or more are eligible to be considered for finance under this scheme.
- *Purpose*: Scheme envisages construction of low cost tanks for collecting and storing rainwater and using it for irrigation, by siphon arrangement, utilizing gravitation flow or by installing motor pump.
- *Amount of Loan*: Maximum amount of finance will be ₹88000/- per acre. Scheme can be adopted in smaller areas also by reducing the cost proportionately.
- *Rate of Interest:* Interest rate ranges from 1.75 per cent below to 2.00 per cent above BPLR for various limits.
- *Repayment*: Repayment based on the income generated from the crops raised and cropping pattern. The maximum period eligible for repayment is 8 years in annual instalments.

AGRICULTURE GOLD LOAN

- *Eligibility*: All individual farmers undertaking cultivation or other activities including allied activities are eligible for short-term finance.
- *Purpose*: To meet genuine credit requirements of farming including allied activities, repairing of equipments and consumption needs etc.
- *Amount of Loan*: The eligible loan amount should be assessed based on the area under cultivation, crops(s) raised, scale of finance and not in relation to the value of gold offered as security.
- *Rate of Interest*: Interest rate ranges from 2.50 per cent below to 1.50 per cent above BPLR for various limits. For working capital loans like ACC/KCC/AGL up to ₹ 3 lakh interest at the rate of 7 per cent is extended as per RBI guidelines subject to the periods stipulated by RBI and beyond that normal rate will apply.
- *Repayment*: As applicable to Agri. Cash Credit accounts depending on the duration of crops raised and harvesting period and income generation, subject to a maximum period of 12 months. The account has to be closed at the end of the repayment period.

SCHEME FOR DEVELOPMENT/ STRENGTHENING OF AGRI. MARKETING INFRASTRUCTURE, GRADING AND STANDARDIZATION

- *Eligibility*: Scheme shall be available to individuals, groups of farmers/ growers/ consumers, partnership/partnership firms, NGO's, SHG, Companies, Corporations, Cooperatives, Co-marketing Federations, Local Bodies, etc.

- *Purpose*: For development of agricultural marketing operations including strengthening of infrastructure, techniques of preservation, storage, etc.
- *Amount of Loan*: As per the project.
- *Rate of Interest*: BPLR irrespective of credit size.
- *Repayment*: Adequate long-term repayment period according to the project.

CONSTRUCTION/ RENOVATION/ EXPANSION OF RURAL GODOWN

- *Eligibility*: The project for construction of rural godowns can be taken up by Individuals, Farmers, Group of farmers/growers, Partnership/ Proprietary firms, NGOs, SHGs, Companies, Corporations, Co-operatives, Federations, Agricultural Produce Marketing Committees, Marketing Boards and Agro Processing Corporations.
- *Purpose*: To create scientific storage capacity with allied facilities in rural areas to meet the requirements of farmers for storing farm produce, processed farm produce and agricultural inputs.
- *Amount of Loan*: As per the project.
- *Rate of Interest*: As applicable to advances under SIB/ CandI segments will be charged.
- *Repayment*: Adequate long-term repayment period, not less than 5 years including a grace period of one year.

CROP LOAN

Crop loan is a short term credit and is generally obtained from primary credit co-op. Society of a village or also from commercial bank. The period of loan is about one year except for sugarcane for which the period is 18 months.

There are two criteria for granting crop loan:

1. One third of gross value
2. Cost of cultivation.

1. One third of gross value approach takes into account the yield and price of the crop, its cost of cultivation and family expenditure. If the gross value is more, more amount of loan becomes available. For example Rice.

	I	II
Yield (Q.)	20	25
Price (₹/Q)	400	400
Gross value (₹.)	8000	10,000
One third (₹.)	2700	3330

2. Thus in second situation farmer is entitled for ₹ 3330 per hectare which is higher than in the first situation. Thus this method takes into account the productive aspect of a crop.

3. In cost of cultivation, direct paid-out costs are only considered. They include items, like seeds, manures, fertilizers, pesticides, diesel/electricity, hired labour etc. In this approach, it is expected that all direct costs to be incurred by the farmer should be covered and accordingly he should get adequate credit. If the cost of all these items of input is ₹ 3500/-. If the loan is granted according to first approach, then the amount which is short, is spent by the farmer from his own funds. Since crop loan is for one season, its recovery is made in one installment after the harvest of the crop. Crop loan is an annual requirement and farmer has to borrow fresh loan for new crop season every time. Therefore, he has to repay the earlier loan with interest within stipulated time. Since this loan is required every season/every year, the procedure of getting this loan is simple and convenient and it is made available by the District Central Co-op. Banks through the village Co-op. Credit Society. So the farmer gets his loan in the village itself. If the loan is to be taken from commercial bank, it is available from the nearby branch of the commercial bank. As for security, the farmer has to offer his land as a security. There is a three tier structure providing crop-loans through co-operative institutions.
   - Appex Bank- State Co-op. Bank.
   - District Central co-op. Bank
   - Village co-op. Credit Society.

Crop-loan is the most important need of the farmer to increase and maintain his productive ability. With the help of this loan amount, he can purchase modern costly inputs and adopt new technologies on his farms. So through these loans co-operative banks play important role in the development and prosperity of agriculture. Among the various types of bank loans to agriculture, the share of crop loan is the highest.

Agricultural Loans

Agricultural loans are available for a multitude of farming purposes. Farmers may apply for loans to buy inputs for the cultivation of food grain crops as well as for horticulture, aquaculture, animal husbandry, floriculture and sericulture businesses. There are also special loans to finance the purchase of agricultural machinery such as tractors, harvesters and trucks. Construction of biogas plants and irrigation systems as well as the purchase of agricultural land may also be financed through special types of agricultural finance.

COOPERATIVE AGRICULTURAL BANK

1. National Bank for Agriculture and Rural Development or NABARD- is responsible for refinance disbursement to commercial banks, State cooperative banks, State cooperatives, rural development banks,

Regional Rural Banks (RRBs) and other eligible financial institutions. It also sanctions money through its Rural Infrastructure Development Fund for projects covering irrigation, rural roads and bridges, health and education, soil conservation and drinking water schemes. NABARD also offers a Kisan Credit Card Scheme and crop loans under the Rashtriya Krishi Bima Yojana.

Banks and RRB's introduced the Kisan Credit Card Scheme of NABARD in their areas of operation. In this scheme eligible farmers are provided with a Kisan Credit Card and a passbook or card-cum-pass book. The revolving cash credit facility allows any number of withdrawals and repayments within the limit. This limit is fixed on the basis of operational land holding, cropping pattern and the scale of finance. Sub-limits may be fixed at the discretion of banks. This Kisan Credit Card is valid for 3 years subject to annual review. As incentive for good performance, credit limits may be enhanced to take care of increase in costs, change in cropping pattern, etc. Each drawl should be repaid within a maximum period of 12 months. Conversion or rescheduling of loans is allowed in case of damage to crops due to natural calamities. Security, margin, rate of interest and other details are fixed according to RBI norms.

2. Bihar State Co-operative Bank Limited (BSCB)- Offers a range of loans and financial schemes to agriculturalists.
3. Haryana State Co-operative Apex Bank Limited (HARCOBANK)- The bank offers crop loans, Kisan Credit Cards, cash credit against hypothecation of stocks and interim finance by way of cash credit.
4. National Federation of State Co-operative Banks Limited (NAFSCOB)- This federation offers a range of agricultural loans through member State Cooperative Banks, District Central Cooperative Banks and Primary Agricultural Cooperative Societies.
5. Orissa State Co-operative Bank Limited (OSCB)- The bank has introduced Kisan Credit Cards in the S.T. Cooperative Credit Sector. It also organizes seminars on agri finance. OSCB has 17 Central Cooperative Banks and around 810 mini banks in different districts of Orissa.
6. Repatriates Co-operative Finance and Development Bank Limited- This bank does not have any specific agricultural loan, but offers a range of financial products that can be accessed by people who wish to develop agriculture and related activities.
7. Punjab State Cooperative Agriculture Development Bank Ltd- Initially, the bank only gave farmers loans to pay off old debts and purchase land. Today, the bank provides loans for various purposes like improvement of alkaline and saline lands, purchase of tractors, installing tube wells and other modern agricultural equipment. It also offers financial schemes for poultry development, dairy

development, horticulture, floriculture, sheep rearing and inland fisheries.

8. Andhra Pradesh State Cooperative Bank Limited (APCOB)- has a loan portfolio that covers crop loans, medium term loans and long term loans for agricultural purposes. It also supports government sponsored District Rural Development Agency projects through IRDP loans and cooperative sugar factories, spinning mills, weaver's societies, employees' cooperative credit societies and other organizations. APCOB has also extended finance to apex cooperative institutions in the State such as APCO, MARKFED and GCC.

NATIONALISED BANKS

1. *Allahabad Bank*: Offers the Kisan Credit Card and Kisan Shakti Yojana Scheme. The Kisan Credit Card is a unique scheme for farmers through which they can draw a cash loan for crop production as well as domestic needs from the card-issuing branch within the sanctioned limit. The Kisan Shakti Yojana provides farm investment credit, as well as personal/domestic loans including repayment of debt to moneylenders. The permissible loan limit will be 50 per cent of the value of land or 5 times the net farm income, whichever is lower, less the outstanding amount, if any, in Agril.
2. *Andhra Bank*: Provides facilities to farmers like AB Kisan Vikas Card, AB Pattabhi Agricard, AB Kisan Chakra, rural godowns, agri clinics, agri service centres, self help groups and solar cookers. They also provide other schemes such as Kisan Sampathi, tractor financing, Kisan Green Card, Surya Sakhti and loans to dairy agents.
3. *Bank of Baroda*: Offers farmers the Baroda Kisan Credit Card. It also has schemes for the purchase of agricultural implements, heavy agricultural machinery like tractors, irrigation and other infrastructure. Bank of Baroda also finances the development of agri industries like horticulture, sericulture, fisheries, dairy and poultry.
4. *Bank of India*: Has a Kisan Credit Card Scheme that helps farmers raise short-term funds for agriculture and other farm-based activities, on an on-going basis, with very flexible and friendly repayment terms. It also offers an agricultural loan for development of agriculture related industries, purchase of machinery and other agricultural purposes.
5. *Bank of Maharashtra*: Offers agriculturists a Maha bank Kisan Credit Card and financial schemes for digging new wells, purchasing harvesters, livestock, vehicles and land. Repayment terms for different agricultural loans range from three to fifteen years.
6. *Canara Bank*: Provides Kisan Credit Cards. Limits up to 50,000 have no margin while those above 50,000 have a margin of 15 to 20 per

cent. Other than this, Canara Bank provides a wide array of financial schemes for different agricultural purposes.

7. *Central Bank of India*: The Central Kisan Credit Card is a credit service provided to farmers on the basis of their holdings for purchasing agricultural inputs. Only those farmers having a good track record for the past 2 years with the bank as a borrower or depositor and who are not defaulters to any credit institution would be considered for loans.
8. *Corporation Bank*: Offers a range of loan schemes to farmers. They are the Corp Gram Mitra Yojana, Corp Arthias Loan Yojana, Corp Kisan Tie-Up Loan Scheme, Corp Kisan Farm Mechanisation Scheme and Corp Kisan Vehicle Loan Yojna.
9. *Dena Bank*: Dena Bank has sponsored 2 Regional Rural Banks namely Dena Gujarat Gramin Bank in Gujarat and Durg Rajnandgaon Gramin Bank (DRGB) in Chhattisgarh. The bank has set up a Rural Development Foundation for training unemployed youth in rural areas. Other financial schemes of the bank are the Dena Swacch Gram Yojana, Dena Kisan Gold Credit Card Scheme and the Dena Bhumiheen Kisan Credit Card Scheme.
10. *Indian Bank*: Has a wide range of schemes for agriculturalists such as Swarojgar Credit Card, Gramin Mahila Sowbhagya Scheme, Kisan Bike Loan Scheme, Yuva KisanVidya Nidhi Yojana and Indian Bank Kisan Card Scheme.
11. *Indian Overseas Bank*: Offers agri business consultancy services that include conducting feasibility and market studies, preparation of detailed project reports and formulation of rehabilitation packages for sick agro units.
12. *Oriental Bank of Commerce*: It has two agricultural projects - the Grameen Project and the Comprehensive Village Development Programme. The Grameen Project involves disbursing small loans ranging from ₹ 75 onwards to mostly women. Training is also provided in villages in using locally available raw material to produce pickles and jams. The Comprehensive Village Development Programme focuses on providing an integrated package of rural finance to villagers to build up their village.
13. *Punjab and Sind Bank*: Offers a range of financial schemes for farmers like the Zimidara Credit Card, tractor finance scheme, drip irrigation scheme, Kheti Udyog Khazana Yojana, vermi composting scheme, horticulture clinic and private veterinary clinic with dairy unit scheme.
14. *Punjab National Bank*: This bank has a special website called PNB Krishi for agriculturalists. It gives details on crop practices, plant protection, farm machinery, market prices and other farming news and activities. The website also provides a list of financial schemes

offered by Punjab National Bank on production credit, investment credit, composite loans, animal husbandry and farm mechanization.

15. *Syndicate Bank*: Offers a wide range of agricultural loan products such as the Synd Jai Kisan Loan Scheme, Jewel Loan Scheme for Agriculture, Syndicate Farm House Scheme, Finance for Hi-tech Agriculture, Development of Irrigation Infrastructure scheme, Syndicate 2/3/4 Wheelers Scheme and the Syndicate Kisan Credit Card (S.K.C.C).
16. *UCO Bank*: This Bank provides the UCO Hirak Jayanti Krishi Yojana to meet the long-term credit needs of the farming community in rural areas for agriculture, allied activities as well as for personal purposes. Only farmers below 60 years are eligible to apply. Minimum quantum of the loan is ₹25,000/- and the maximum is ₹5 lakhs.
17. *Union Bank of India*: Facilities provided to farmers include Kisan ATM Cards and special Kisan ATM Machines. These ATM's are easy to operate and do not require farmers to have a high level of literacy. They are voice enabled in the local language, have a touch screen monitor and work on a bio-metric authentication system like finger print verification.
18. *United Bank of India*: The range of financial schemes offered to agriculturalists include the United Krishi Laghu Paribahan Yojana, United Krishi Sahayak Yojana, United Gramya shree Yojana, Gramin Bhandaran Yojana and the United Bhumiheen Kisan Credit Card.
19. *Vijaya Bank*: This bank offers one comprehensive financial scheme known as the Vijaya Krishi Vikas (VKV) Scheme. This scheme provides a simple package to farmers to meet entire agricultural credit requirements such as crop production, investment credit and consumption credit. All farmers including owners, tenant cultivators, leased land farmers and sharecroppers are eligible for this scheme.

FINANCING OF THE FOOD PROCESSING SECTOR

The structural complexion of the Indian supply chain has led to limited scale of financing as well as higher risk, given the lack of control of each of the players, on the supply chain.

Financing of farmers: Farmers often rely on unorganized sources of credit due to bottlenecks in access, timeliness in availability and adequacy of credit from organized sources. The key hurdles faced by banks in financing farmers are their inability to provide adequate collateral as security, and the potential for default, in the absence of an assured market for their produce.

Financing of food processing enterprises: The food processing enterprises primarily comprise small and medium sized companies, a large proportion of which have stand alone operations, with no linkages with farmers, and reliant on other organizations to undertake marketing/further processing of their products. Consequently, companies in the food processing sector usually bear a steep cost of interest for the high risk perception associated with the nature of their operations.

Further there are several regulations which limit availability of finance to the sector, including the definition of priority sector lending, the Cooperatives Act and the Warehousing Corporation Act.

Absence of well-developed risk mitigation tools has further impacted availability of finance to the food and agriculture sector. Additionally, crop insurance schemes in India have faced various issues including assessment of farmer yields at the mandal level, lack of past statistical data to calculate premia, high premia for certain crops and administrative hurdles in managing claims, thus resulting in poor economic viability as also limited benefit to farmers.

FDI in food processing: FDI in the food processing sector is low, constituting a mere 4 per cent of total FDI in the period 1991-2004.The actual inflow is only about 28 per cent of FDI approved (INR 4.2 bn out of approved INR 11.6 bn) This is despite the fact that with the exception of food retailing, plantations and alcoholic beverages, 100 per cent FDI is permitted in the sector.

Government Schemes: Government provides assistance to the industry and entrepreneurs under various schemes. While there are significant overlaps in these on the one hand, there are several unaddressed need gaps of the industry, on the other. Further, the restriction on quantum of financial assistance per unit disincentivises scale. Additionally, the structure of the scheme needs to be based on specific sectoral requirements. The Government needs to have a robust monitoring mechanism to assess the direct and indirect impact of such assistance.

Market Intelligence

Unlike in several leading food producing countries in the world, there is no updated and reliable source of information pertaining to the food processing sector such as on government policies, demand supply outlook, food standards etc. The process of obtaining information from existing industry agencies/bodies, even if available, is tedious. Further, there is limited sharing of information across institutes/bodies collating data. Small and medium sized players are particularly at a disadvantage on account of lack of availability of information.

Food Safety, Hygiene and Regulations

Food safety is a growing concern across the world. There is an increasing need to provide greater assurance about the safety and quality of food to consumers. Compliance with international food standards is a prerequisite to gain a higher share of world trade. At the same time there is growing consciousness among Indian consumers given recent controversies on quality standards of many food products. The existing mechanism for setting and harmonizing food standards is inadequate in the context of the increasing importance of food hygiene and safety and evolution of international standards. This is due to limited infrastructure for generating/analysing scientific data and food testing. Many laboratories do not have basic facilities to test antibiotic residues, heavy metal contamination and other toxic contaminants in food products. Testing manuals do not describe parameters and procedures adequately. Moreover, limited coordination between the various food testing laboratories has led to inefficient utilization of the food testing infrastructure available in the country. Further, Indian food regulations comprise various food laws which are under the ambit of various ministries, which has lead to operational complexities for processors.

Taxation

The tax incidence (excise/customs, state taxes) in Indian food products is higher than many other countries. This has led to low affordability of processed products. Concessions have been extended to different products of food processing industry in the last few budgets but it happened in a selective manner. The reasons based on which concessions were extended for one category hold true for all sectors of the industry.

Apart from the above, there are several other sector-specific issues, which influence the growth of the food sector. These include lack of cultivation of varieties amenable to processing, unequal advantage to the unorganized sector, absence of direct farmer-processor linkages and lack of focus on tapping specific sectoral advantages.

Research and Development

The Indian food industry needs world class technologies to compete effectively with international products. Indian R&D institutions have not been able to develop innovative products, processes and machinery of global stature as reflected in India's share in global trade. The industry is dissatisfied with the quality of research output and response of existing R&D institutions. The key reasons for this are segregation of academics from applied research, inadequate industry interface, low commercial orientation and lack of collaborative efforts with global peers.

FINANCE FOR AGRICULTURE & FOOD PROCESSING

Banks: The Food Processing sector has access to credit from Commercial Banks-Indian and Foreign, Cooperative Banks and the Regional Rural Banks. The formal institutional sector assists the food processing companies through long term loans for capital investments and short term loans for working capital. The total lending to the food processing sector is clubbed with lending to the agriculture sector and separate details for financing to the food processing sector are not available.

- National Bank for Agricultural and Rural Development (NABARD): NABARD offers refinance facilities for food processing, agri infrastructure, developmental assistance for RRBs, DCCBs, SSI and

SHG linkages, and assists in infrastructure development and research.

- Small Industries Development Bank of India (SIDBI):SIDBI has been involved in assisting the entire SSI Sector including tiny, village and cottage industries through suitable schemes tailored to address the funding requirements for expansion, diversification, modernisation and rehabilitation.
- Export Import Bank (EXIM Bank): Exim Bank assists in financing and facilitation of foreign trade. The Bank also offers financial support to companies engaged in exports.
- National Cooperative Development Corporation (NCDC): NCDC assists in promotion, planning and financing the agricultural supply chain from production, processing, storage and trade of agricultural produce and food products. NCDC also provides assistance for marketing of certain notified commodities *e.g.* fertilizers, pesticides, agricultural machinery etc.
- Ministries/Government bodies
- Ministry of Food Processing Industries (MFPI): MFPI is the Nodal agency for development of the processed food sector in the country MFPI's financial schemes include schemes for technology upgradation, Human resource development, Quality testing, R&D,TQM, backward and forward integration, development of infrastructure including food parks.
- Agricultural and Processed Foods Products Export Development Authority (APEDA): APEDA facilitates market linkages between Indian producers, manufacturers and the international market. APEDA provides financial assistance for market development, infrastructure development and development of quality enhancing facilities.
- Ministry of Agriculture (MoA), Government of India: The Ministry of Agriculture under various schemes, provides financial assistance for development of specific crops for investment in seeds, irrigation, farm implements, inputs, infrastructure and training.
- National Horticultural Board (NHB):NHB promotes integrated development in horticulture, assists in development of post harvest management infrastructure, promotes production and processing of fruits and vegetables, strengthening of market information systems and assists in R&D programmes in cultivation and processing. NHB's financial schemes are directed towards commercial horticulture and infrastructure related to post harvest techniques.

Financial assistance from these organizations are in the form of grants, back-ended subsidies, soft loans, refinance etc., with most of the schemes directed to specific sub-sectors of the agri/food processing industry.

In addition, some State Governments offer financial schemes to companies in the sector. The key issues with Government schemes are highlighted below:

- Overlap among schemes: Various Government agencies have separate schemes which aim to achieve the same objectives. Thus, an applicant can potentially approach different agencies and obtain assistance from more than one agency. Some of the examples of overlapping schemes are as follows:
- Several areas addressed inadequately: There are a host of areas which are not addressed by these schemes such as financing of freezer cabinets for retail outlets, financing of bulkcoolers for milk, financing of alternate markets/mandis, financing of vending machines for tea/coffee/beverages etc.
- Creation of excess capacity: There are instances where assistance has been provided to set up cold storages in locations which have excess storage capacity. This has lead to an unviable situation for all cold storage operators in the region. It is essential for the nodal agencies to achieve development of the food processing sector in a sustainable manner.
- Inadequate financial assistance^ most instances, there is a cap on total financial assistance provided at INR 5-5.5 mn. This leads to fragmentation in capacity creation, and often, a situation of overcapacity. At the same time, it does not allow players to scale up their operations.
- Inadequate Monitoring of progress of schemes: It is important to track progress of the projects funded by the Government. Periodic appraisal of the progress will ensure that funds are appropriately utilized. This will also enable assessment of success and lacunae of these schemes and record the multiplier effect of the financial support provided.
- Most of the schemes are back-ended which leads to a funds crunch during implementation.
- Institutions providing subsidies do not undertake their own assessment and rely on the appraisal report of the bank which provides loans to the project.
- There are significant time lags from the date of application for financial assistance, to release of funds, and may affect the project schedule, and lead to cost overruns
- These schemes do not address working capital requirements, which due to the nature of the industry are significant and often higher in terms of quantum than capital investment requirements
- Applicants claim that they are often unaware that their applications have been rejected on account of lack of communication from the

funding institution. Further, applicants are unaware of the reasons for rejection.

SOURCES OF FINANCE FOR AGRICULTURE

At present nearly 60 per cent of rural credit is used for agriculture and allied activities, 10 per cent for non-farm activities and the balance 30 per cent for funding household consumption expenditure. Nearly 40-45 per cent of rural credit needs are catered to by formal credit institutions and the balance by the informal sector including commission agents, input suppliers, traders, Self Help Groups (Joint Liability Groups), processing industries and professional money lenders.

The Indian commercial banks are mandated by the Reserve Bank of India (RBI), to undertake directed lending for agriculture and rural development. Commercial banks are required to achieve priority sector lending of 40 per cent of net bank credit of which 18 per cent should be for agriculture. Further, sub-targets are also specified *i.e.,* Direct Agriculture (credit for direct farmer benefit, Target: Minimum of 13.5 per cent of the net bank credit) and Indirect Agriculture (Target: Maximum of 4.5 per cent of the net bank credit).

In FY 03, total agricultural advances outstanding under priority sector lending, by Public Sector Banks stood at INR 735 billion, on a total net bank credit portfolio of INR 4779 billion. As compared to this, agricultural advances by Private Sector Banks stood at INR 119 billion, on a total net bank credit portfolio of INR 718 billion. The following Exhibit provides details of direct and indirect priority sector credit amongst Public and Private sector banks during FY03.

In case of a shortfall in lending to agriculture, banks are required to invest in Rural Infrastructure Development Fund (RIDF) deposits, which offers interest rates linked to the bank's performance in lending to agriculture. These rates are inversely proportional to the shortfall in agricultural lending. The interest rate on RIDF is provided in slabs, based on the achievement of the priority sector lending target, and can be as low as 4 per cent to 5 per cent.Therefore, it is imperative to achieve the agriculture target and more specifically the Direct Agriculture target as it accounts for 75 per cent of the total. Established in 1996, RIDF is being utilized for making investments in rural infrastructure projects. Nine tranches of RIDF have been established with an aggregate corpus of INR 340 billion, with the ninth tranche of INR 55 billion. Cumulative sanctions and disbursements under various tranches of RIDF stood at INR 334 billion and INR 193 billion respectively, at the end of February 2004.The RIDF funds are used to finance minor irrigation projects, shallow wells, drinking water facilities, roads, bridges and storage infrastructure.

FINANCING-FOOD PROCESSING SECTOR

The food processing sector comprises a large number of small and medium sized companies, a significant proportion of which have stand alone operations, with no control over the raw material base and reliant on other organizations to undertake marketing/further processing of their products. Banks and financial Institutions adopt the same risk models relevant to the manufacturing sector, for assessing food processing companies. Interest charges for food processing companies are high, on account of the high risk perception associated with the nature of their operations.

FOOD PROCESSORS' PERSPECTIVE

With raw material availability often being seasonal, and/or concentration in demand, inventory holdings are high and thus the working capital finance issued through normal Maximum Permissible Bank Finance method does not address the funding requirements adequately. The rate of interest for working capital assistance is high. Most food processing companies are not able to access adequate working capital at reasonable rates. This affects the raw material procurement and capacity utilization through out the year.

Unavailability of large working capital facilities, or facilities without a high seasonal drawal, restricts players in purchasing agri produce in large volumes, when prices are favourable.

Banker's Perspective

- Lack of reliable information on demand-supply, price trends, raw material supplies etc is a significant constraint for banks, impacting credit assessment and monitoring
- High operational/transaction cost in servicing small companies
- Lack of backward linkages for assured access to raw material
- Lack of forward linkages of stand-alone food processing units for marketing/distribution

Owing to high cost of market development, the profitability of food processing companies is under pressure in the initial years of their operation. This impacts the risk rating and leads to high cost of borrowing. Even temporary adversities in market conditions can lead to defaults and eventually the loan can become a non performing asset for banks.

2

Technical Processing in Agricultural Production

Above we have already had occasion to note that writers on agriculture, in classifying systems of farming according to the principal market product, assign the industrial or technical system of farming to a special category. The essence of this system is that the agricultural product, before going into consumption (personal or productive), undergoes technical processing. The establishments which effect this processing either constitute part of the very farms on which the raw material is produced or belong to special industrialists who buy up the raw material from the peasant farmers. From the standpoint of political economy the difference between these two types is unimportant. The growth of agricultural technical trades is extremely important as regards the development of capitalism. Firstly, this growth represents one of the forms of the development of commercial farming, and is, moreover, the form that shows most vividly the conversion of agriculture into a branch of industry of capitalist society.

Secondly, the development of the technical processing of agricultural produce is usually connected intimately with technical progress in agriculture: on the one hand, the very production of the raw material for processing often necessitates agricultural improvement (the planting of root-crops, for example); on the other hand, the waste products of the processing are frequently utilised in agriculture, thus increasing its effectiveness and restoring, at least in some measure, the equilibrium, the interdependence, between agriculture and industry, the disturbance of which constitutes one of the most profound contradictions of capitalism. We must accordingly now describe the development of technical agricultural trades in post-Reform Russia.

DISTILLING

Here we regard distilling only from the point of view of agriculture. Accordingly, there is no need for us to dwell on the rapid concentration of distilling in large plants (partly due to excise requirements), on the rapid progress of factory technique, with the consequent cheapening of production, and the increase in excise duties which has outstripped this cheapening of production and because of its excessive amount has retarded the growth of consumption and production.

Thus, over 9/10 of the distilleries (accounting for over 4/5 of the total output) are directly connected with agriculture. Being large capitalist enterprises, these establishments lend the same character to all the landlord farms on which they are set up (the distilleries belong almost without exception to landlords, mainly to members of the nobility). The type of commercial farming under review is particularly developed in the central black-earth gubernias, in which are concentrated over 1/10 of the total number of distilleries in the Russian Empire (239 in 1896-97, of which 225 were agricultural and mixed), producing over a quarter of the total output of spirits (7,785,000 vedros in 1896-97, of which 6,828,000 at agricultural and at mixed distilleries).

Thus in the area where labour-service predominates, the commercial character of agriculture most frequently (as compared with other areas) manifests itself in the distilling of vodka from grain and potatoes. Distilling from potatoes has undergone a particularly rapid development since the Reform, as may be seen from the following data relating to the whole of the Russian Empire:

Thus, with a general twofold increase in the quantity of crops distilled, the quantity of potatoes used increased about 15-fold. This fact strikingly corroborates the proposition established above that the enormous increase in the potato area and crop signifies the growth of precisely commercial and capitalist farming, along with improvement of agricultural technique, with the replacement of the three-field system by multi-field crop rotation, etc. The area of the biggest development of distilling is also distinguished for the biggest (in the Russian gubernias, *i.e.*, not counting the Baltic and the western gubernias) net per-capita harvest of potatoes. Thus in the northern black-earth gubernias the figures for 1864-1866, 1870-1879 and 1883-1887 were 0.44, 0.62 and 0.60 chetverts respectively, whereas for the whole of European Russia (50 gubernias) the corresponding figures were 0.27, 0.43 and 0.44 chetverts. As far back as the beginning of the 80s the *Historico-Statistical Survey* noted that "the region marked by the greatest expansion of potato cultivation covers all the gubernias of the central and northern parts of the black-earth belt, the Volga and Transvolga gubernias and the central non-black-earth gubernias" (*loc. cit.*, p. 44).

The expansion of potato cultivation by landlords and well-to-do peasants means an increase in the demand for hired labour; the cultivation of a dessiatine of potatoes absorbs much more labour than the cultivation of a dessiatine of cereals and the use of machinery in, for example, the central black-earth area is still very slight. Thus, while the number of workers engaged in the distilling industry proper has decreased, the elimination of labour-service by the capitalist system of farming, with the cultivation of root-crops, has increased the demand for rural day labourers.

BEET-SUGAR PRODUCTION

The processing of sugar-beet is even more highly concentrated in big capitalist enterprises than distilling is, and is likewise an adjunct of the landlords' (mainly noblemen's) estates. The principal area of this industry is the south-western gubernias, and then the southern black-earth and central black-earth gubernias. The area under sugar-beet amounted in the 60s to about 100,000 dess., in the 70s to about 160,000 dess.; in 1886-1895 to 239,000 dess., in 1896-1898 to 369,000 dess., in 1900 to 478,778 dess., in 1901 to 528,076 dess. (*Torgovo-Promyshlennaya Gazeta,* 1901, No. 123), in 1905-06 to 483,272 dess. (*Vestnik Finansov,* 1906, No. 12).

Hence, in the period following the Reform the area cultivated has increased more than 5-fold. Incomparably more rapid has been the growth of the amount of sugar-beet harvested and processed: on an average the weight of sugar-beet processed in the Empire in the years 1860-1864 was 4.1 million berkovets; in 1870-1874—9.3 million; in 1875-1879—12.8 million; in 1890-1894—29.3 million; and in 1895-96 and 1897-98—35 million. The amount of processed sugar-beet has grown since the 60s more than 8-fold. Hence, there has been an enormous increase in the beet yield, *i.e.,* in labour productivity, on the big estates organised on capitalist lines.

The introduction of a root-plant like beet into the rotation is indissolubly linked with the transition to a more advanced system of cultivation, with improved tillage and cattle feed, etc. "The tillage of the soil for beetroot," we read in the *Historico-Statistical Survey,* "which, generally speaking, is rather complicated and difficult, has been brought to a high degree of perfection on many beet farms, especially in the south-western and Vistula gubernias. In different localities, various more or less improved implements and ploughs are used for tilling; in some cases even steam ploughing has been introduced".

This progress of large-scale capitalist farming gives rise to quite a considerable increase in the demand for agricultural wage-workers— regular and particularly day labourers—the employment of female and child labour being particularly extensive. Among the peasants of the neighbouring gubernias a special type of migration has arisen, known as migration "to sugar". It is estimated that the complete cultivation of a morg (= 2/3 dess.) of beet land requires 40 working days. The *Combined Material on the Position of the Rural Population* (published by Committee of Ministers) estimates that the cultivation of one dessiatine of beet land, when done by machine, requires 12, and when by hand 25, working days of males, not counting women and juveniles. Thus, the cultivation of the total beet area in Russia probably engages not less than 300,000 agricultural day labourers, men and women. But the increase in the number of dessiatines under beet is not enough to give a complete idea of the demand for hired labour, since some jobs are paid for at so much per berkovets. Here, for example, is what we read in *Reports and Investigations of Handicraft Industry in Russia.*

"The female population both of the town, and of the uyezd" "think highly of work on the beet fields; in the autumn the cleaning of beets is paid at 10 kopeks per berkovets, and two women clean from six to ten berkovets a day, but some contract to work during the growing season as well, weeding and hoeing; in that case, for the full job, including digging and cleaning, they get 25 kopeks per berkovets of cleaned beets." The conditions of the workers on the beet plantations are extremely bad. For instance, the *Vrachebnaya Khronika Kharkovskoi Gubernii* cites "a number of exceedingly deplorable facts about the conditions *of those working on the red-beet plantations*. Thus, the Zemstvo physician, Dr. Podolsky, of the village of Kotelva, Akhtyrka Uyezd, writes: 'In the autumn *typhus* usually breaks out among *young people employed on the red-beet plantations of the well-to-do peasants*.

The sheds assigned for the workers' leisure and sleeping quarters are kept by such planters in a very filthy condition; by the time the job ends the straw used for sleeping is literally converted into dung, for it is never changed: this becomes a breeding ground of infection. Typhus has had to be diagnosed immediately in the case of four or five patients brought in from one and the same plantation.' In the opinion of this doctor, 'most of the syphilis cases come from the red-beet plantations.' Mr. Feinberg rightly asserts that 'work on the plantations, which is no less injurious to the workers themselves and to the surrounding population than work in the factories, has particularly disastrous consequences, because large numbers of women and juveniles are engaged in it, and because the workers here are without the most elementary protection from society and the State'; in view of this, the author wholly supports the opinion expressed by Dr. Romanenko at the Seventh Congress of Doctors of Kharkov Gubernia that 'in issuing *compulsory* regulations, *consideration* must also be given *to the conditions of the workers on the beet plantations*.

These workers lack the most essential things; they live for months under the open sky and eat from a common bowl.'" Thus, the growth of beet cultivation has enormously increased the demand for rural workers, converting the neighbouring peasantry into a rural proletariat. The increase in the number of rural workers has been but slightly checked by the inconsiderable drop in the number of workers engaged in the beet-sugar industry proper.

POTATO-STARCH PRODUCTION

From branches of technical production conducted exclusively on landlord farms let us pass to such as are more or less within the reach of the peasantry. These include, primarily, the processing of potatoes (partly also wheat and other cereals) into starch and treacle. Starch production has developed with particular rapidity in the post-Reform period owing to the enormous growth of the textile industry, which raises a demand for starch. The area covered by this branch of production is mainly the non-black-earth, the industrial, and, partly, the northern black-earth gubernias.

The *Historico-Statistical Survey* estimates that in the middle of the 60s, there were about 60 establishments with an output valued at about 270,000 rubles, while in 1880 there were 224 establishments with an output valued at 1,317,000 rubles. In 1890, according to the *Directory of Factories and Works* there were 192 establishments employing 3,418 workers, with an output valued at 1,760,000 rubles.

"In the past 25 years," we read in the *Historico-Statistical Survey*, "the number of establishments engaged in starch production has increased 4 1/2 times and the total output 10 3/4 times; nevertheless, this productivity is far from covering the demand for starch", as evidenced by the increased starch imports from abroad. Analysing the data for each gubernia, the *Historico-Statistical Survey* reaches the conclusion that our production of potato starch (unlike that of wheat-starch) is of an agricultural character, being concentrated in the hands of peasants and landlords. "Showing promise of extensive development" in the future, "it is even now furnishing our rural population with considerable advantages".

We shall see in a moment who enjoys these advantages. But first let us note that two processes must be distinguished in the development of starch production: on the one hand, the appearance of new small factories and the growth of peasant production, and on the other, the concentration of production in large steam-powered factories. For instance, in 1890 there were 77 steam-powered factories, with 52 per cent of the total number of workers and 60 per cent of the total output concentrated in them. Of these works only 11 were established before 1870, 17 in the 70s, 45 in the 80s, and 2 in 1890.

To acquaint ourselves with the economy of peasant starch production, let us turn to local investigations. In Moscow Gubernia, in 1880-81, 43 villages in 4 uyezds engaged in starch making. The number of establishments was estimated at 130, employing 780 workers and having an output valued at not less than 137,000 rubles. The industry spread mainly after the Reform; its technique gradually improved and larger establishments were formed requiring more fixed capital and showing a higher productivity of labour. Hand graters were replaced by improved ones, then horse power appeared, and finally the *drum* was introduced, considerably improving and cheapening production. Here are data we have compiled from a house-to-house census of "handicraftsmen," according to size of establishment: Thus we have here small capitalist establishments in which, as production expands, the employment of hired labour increases and the productivity of labour rises. These establishments bring the peasant bourgeoisie considerable profit, and also improve agricultural technique. But the situation of the workers in these workshops is very unsatisfactory, owing to the extremely insanitary working conditions and the long working day.

The peasants who own "grating" establishments farm under very favourable conditions. The planting of potatoes (on allotment, and chiefly on rented land) yields a considerably larger income than the planting of rye or oats. To enlarge their business the workshop owners rent a considerable amount of allotment land from the poor peasants. For example, in the village of Tsybino (Bronnitsy Uyezd), 18 owners of starch workshops (out of 105 peasant families in the village) rent allotments from peasants who have left in search of employment, and also from horseless peasants, thus adding to their own 61 allotments 133 more, which they have rented; concentrated in their hands are a total of 194 allotments, *i.e.*, 44.5 per cent of the total number of allotments in the village.

"Exactly similar things," we read in the *Returns*, "are met with in other villages where the starch industry is more or less developed" (*loc. cit.*, 42). The owners of the starch workshops have twice as much livestock as the other peasants: they average 3.5 horses and 3.4 cows per household, as against 1.5 horses and 1.7 cows among the local peasants in general. Of the 68 workshop owners (covered by the house-to-house census) 10 own purchased land, 22 rent non-allotment land and 23 rent allotment land. In short, these are typical representatives of the peasant bourgeoisie.

Exactly analogous relations are to be found in the starch making industry in the Yuryev Uyezd, Vladimir Gubernia. Here, too, the workshop owners carry on production mainly with the aid of wage-labour (out of 128 workers in 30 workshops, 86 are hired); and here, too, the workshop owners are far above the mass of the peasantry as far as stock-breeding and agriculture are concerned; they use potato pulp as feed for their cattle. Even real capitalist farmers emerge from among the peasants. Mr. Prugavin describes the farm of a peasant who owns a starch works (valued at about 1,500 rubles) employing 12 wage-workers. This peasant grows potatoes on his own farm, which he has enlarged by renting land. The crop rotation is seven-field and includes clover. For the farm work he employs from 7 to 8 workers, hired from spring to autumn ("from end to end"). The pulp is used as cattle feed, and the owner intends to use the waste water for his fields.

Mr. V. Prugavin assures us that this works enjoys "quite exceptional conditions." Of course, in any capitalist society the rural bourgeoisie will always constitute a very small minority of the rural population, and in this sense will, if you like, be an "exception." But this term will not eliminate the fact that in the starch-making area, as in all the other commercial farming areas in Russia, a class of rural entrepreneurs is being formed, who are organising capitalist agriculture.

VEGETABLE-OIL PRODUCTION

The extraction of oil from linseed, hemp, sunflower and other seeds is also frequently an agricultural industry. One can gauge the development of vegetable-oil production in the post-Reform period from the fact that in 1864 the vegetable-oil output had an estimated value of 1,619,000 rubles, in 1879 of 6,486,000 rubles, and in 1890 of 12,232,000 rubles. In this branch of production, too, a double process of development is to be observed: on the one hand, small peasant (and sometimes also landlord) oil presses producing oil for sale are established in the villages. On the other hand, large steam-driven works develop, which concentrate production and oust the small establishments. Here we are interested solely in the agricultural processing of oil-bearing plants. "The owners of the hempseed oil presses," we read in the *Historico-Statistical Survey*, "belong to the well-to-do members of the peasantry"; they attach particular value to vegetable-oil production because it enables them to obtain excellent feed for their cattle (oil cake).

Mr. Prugavin (*loc. cit.*), noting the "extensive development of the production of linseed oil" in the Yuryev Uyezd, Vladimir Gubernia, states that the peasants derive "no little advantage" from it, that crop and stock raising is conducted on a far higher level by peasants who own oil presses than by the bulk of the peasantry and that some of the oil millers also resort to the hire of rural workers. The Perm handicraft census for 1894-95 also showed that crop raising is conducted on a much higher level by handicraft oil millers than by the bulk of the peasants (larger areas under crops, far more animals, better harvests, etc.), and that this improvement in cultivation is accompanied by the hiring of rural workers. In the post-Reform period in Voronezh Gubernia, there has been a particular development of the commercial cultivation of sunflower seed, which is crushed for oil in local presses.

The area under sunflowers in Russia in the 70s was estimated at about 80,000 dess. (*Historico-Statistical Survey*, I), and in the 80s at about 136,000 dess., of which 2/3 belonged to peasants. "Since then, however, judging by certain data, the area under this plant has considerably increased, in some places by 100 per cent and even more" (*Productive Forces*, I, 37). "In the village of Alexeyevka alone" (Biryuch Uyezd, Voronezh Gubernia), we read in the *Historico-Statistical Survey*, Vol. II, "there are more than 40 oil presses, and Alexeyevka itself, solely owing to sunflowers, has prospered and grown from a wretched little hamlet into a rich town ship, with houses and shops roofed with sheet iron" (p. 41). How this wealth of the peasant bourgeoisie was reflected in the condition of the mass of the peasantry may be seen from the fact that in 1890, in the village of Alexeyevka, out of 2,273 families registered (13,386 persons of both sexes), 1,761 had no draught animals, 1,699 had no implements, 1,480 cultivated no land, and only 33 families did not engage in industries.

TOBACCO GROWING

In conclusion, let us make some brief observations on the development of tobacco growing. The average crop in Russia for the years 1863-1867 was 1,923,000 poods from 32,161 dess.; for 1872-1878 it was 2,783,000 poods from 46,425 dess.; for the 80s, it was 4 million poods from 50,000 dess. The number of plantations in the same periods was estimated at 75,000, 95,000 and 650,000 respectively, which evidently indicates a very considerable increase in the number of small cultivators drawn into this type of commercial farming. Tobacco growing requires a considerable number of workers. Among the types of agricultural migration note is therefore made of migration to tobacco plantations (particularly to the outer gubernias in the South, where the cultivation of tobacco has recently expanded with exceptional rapidity). Reference has already been made in publications to the fact that the workers on the tobacco plantations lead a very hard life.

In the *Survey of Tobacco Growing in Russia,* there are very detailed and interesting data on tobacco growing as a branch of commercial farming. Mr. V. S. Shcherbachov, describing tobacco growing in Malorossia, gives wonderfully precise information on three uyezds of Poltava Gubernia (Priluki, Lokhvitsa and Romny). This information, gathered by the author and arranged by the Bureau of Statistics, Poltava Gubernia Zemstvo Board, covers 25,089 peasant farms in the three uyezds that grow tobacco; they have 6,844 dessiatines under tobacco and 146,774 dessiatines under cereals. We see an enormous concentration of both the tobacco and the cereal area in the hands of the capitalist farms. Less than one-eighth of the farms (3,000 out of 25,000) hold more than half the area under cereals (74,000 dess. out of 147,000), with an average of nearly 25 dess. per farm. Almost half the area under tobacco (3,200 dess. out of 6,800) belongs to these farms, the average per farm being over 1 dessiatine, whereas for all the other groups the area under tobacco does not exceed one-to two-tenths of a dessiatine per household.

The owner of 7 dessiatines under tobacco must there fore have at least 14 workers; in other words, he must undoubtedly base his farm on wage-labour. Some grades of tobacco require not two but three seasonal workers per dessiatine, and day labourers in addition. In a word, we see quite clearly that the greater the degree to which agriculture becomes commercial, the more highly developed is its capitalist organisation.

The preponderance of small and tiny farms among the tobacco growers (11,997 farms out of 25,089 have *up to one-tenth* of a dessiatine planted) does not in the least refute the fact of the capitalist organisation of this branch of commercial agriculture; for this mass of tiny farms accounts for an insignificant share of the output (11,997, *i.e.*, nearly half the farms, have in all 522 dess. out of 6,844, or less than one-tenth). Nor do "average" figures, to which people so often confine themselves, provide a picture of the real situation (the average per farm is a little over 114 dessiatine under tobacco).

In some uyezds the development of capitalist agriculture and the concentration of production are still more marked. In the Lokhvitsa Uyezd, for example, 229 farms out of 5,957 each have 20 dessiatines and more under cereals. Their owners have 22,799 dess. under cereals out of a total of 44,751, *i.e.*, more than half. Each farmer has about 100 dess. under crops. Of the land under tobacco they have 1,126 dess. out of 2,003 dess. And if the farms are grouped according to area under tobacco, we have in this uyezd 132 farmers out of 5,957 with two and more dessiatines under tobacco. These 132 farmers have 1,441 dess. under tobacco out of 2,003, *i.e.*, 72 per cent and more than *ten* dessiatines under tobacco per farm. At the other extreme of the same Lokhvitsa Uyezd we have 4,360 farms (out of 5,957) having up to one-tenth of a dessiatine each under tobacco, and altogether 133 dessiatines out of 2,003, *i.e.*, 6 per cent. It goes without saying that the capitalist organisation of *production* is accompanied here by a very considerable development of *merchant's* capital and by all sorts of exploitation outside the sphere of production. The small tobacco growers have no drying sheds, are unable to give their tobacco time to ferment and to sell it (in 3 to 6 weeks) as a finished product. They sell the unfinished product at *half the price* to buyers-up, who very often plant tobacco themselves on rented land. The buyers-up "squeeze the small planters in every way" (p. 31 of cited publication). Commercial agriculture is commercial capitalist production: this relation can be clearly traced (if only one is able to select the proper methods) in this branch of agriculture too.

AGRICULTURAL SURPLUSES

Patrick Mulvany, Chairman of the UK Food Group, accused some governments, especially the Bush administration, of using GM food aid as a way to dispose of unwanted agricultural surpluses. The UN blamed food companies and accused them of violating human rights, calling on governments to regulate these profit-driven firms. It is true that the acceptance of biotechnology and genetically modified foods will also benefit rich research companies and could possibly benefit them more than consumers in underdeveloped nations.

LABELLING

While some groups advocate the complete prohibition of GMOs, others call for mandatory labelling of genetically modified food or other products. Other controversies include the definition of patent and property pertaining to products of genetic engineering.

UNDERDEVELOPED NATIONS

Some groups believe that underdeveloped nations will not reap the benefits of biotechnology because they do not have easy access to these developments, cannot afford modern agricultural equipment, and certain aspects of the system revolving around intellectual property rights are unfair to undeveloped countries. For example, The CGIAR (Consultative Group of International Agricultural Research) is an aid and research organization that has been working to achieve sustainable food security and decrease poverty in undeveloped countries since its formation in 1971.

In an evaluation of CGIAR, the World Bank praised its efforts but suggested a shift to genetics research and productivity enhancement. This plan has several obstacles such as patents, commercial licenses, and the difficulty that third world countries have in accessing the international collection of genetic resources and other intellectual property rights that would educate them about modern technology. The International Treaty on Plant Genetic Resources for Food and Agriculture has attempted to remedy this problem, but results have been inconsistent. As a result, "orphan crops", such as tef, millets, cowpeas, and indigenous plants, are important in the countries where they are grown, but receive little investment.

PRIVATE INVESTMENTS

The development and implementation of policies designed to encourage private investments in research and marketing biotechnology that will meet the needs of poverty-stricken nations, increased research on other problems faced by poor nations, and joint efforts by the public and private sectors to ensure the efficient use of technology developed by industrialized nations have been suggested. In addition, industrialized nations have not tested GM technology on tropical plants, focusing on those that grow in temperate climates, even though undeveloped nations and the people that need the extra food live primarily in tropical climates. Many European scientists are disturbed by the fact that political factors and ideology prevent unbiased assessment of the GM technology in some EU countries, with a negative effect on the whole community.

TRANSGENIC ORGANISMS

Another important controversy is the possibility of unforeseen local and global effects as a result of transgenic organisms proliferating. The basic ethical issues involved in genetic research are discussed in the article on genetic engineering.

Some critics have raised the concern that conventionally-bred crop plants can be cross-pollinated (bred) from the pollen of modified plants. Pollen can be dispersed over large areas by wind, animals and insects. In 2007, the U.S. Department of Agriculture fined Scotts Miracle-Gro $500,000 when modified genetic material from creeping bentgrass, a new golf-course grass Scotts had been testing, was found within close relatives of the same genus (*Agrostis*) as well as in native grasses up to 21 km (13 miles) away from the test sites, released when freshly cut grass was blown by the wind.

GM proponents point out that outcrossing, as this process is known, is not new. The same thing happens with any new open-pollinated crop variety—newly introduced traits can potentially cross out into neighbouring crop plants of the same species and, in some cases, to closely related wild relatives. Defenders of GM technology point out that each GM crop is assessed on a case-by-case basis to determine if there is any risk associated with the outcrossing of the GM trait into wild plant populations.

The fact that a GM plant may outcross with a related wild relative is not, in itself, a risk unless such an occurrence has negative consequences. If, for example, a herbicide resistance trait was to cross into a wild relative of a crop plant it can be predicted that this would not have any consequences except in areas where herbicides are sprayed, such as a farm. In such a setting the farmer can manage this risk by rotating herbicides. The European Union funds research programmes such as Co-Extra, that investigate options and technologies on the coexistence of GM and conventional farming. This also includes research on biological containment strategies and other measures, to prevent outcrossing and enable the implementation of coexistence.

If patented genes are outcrossed, even accidentally, to other commercial fields and a person deliberately selects the outcrossed plants for subsequent planting then the patent holder has the right to control the use of those crops. This was supported in Canadian law in the case of Monsanto Canada Inc. v. Schmeiser.

"TERMINATOR" AND "TRAITOR"

An often cited controversy is a "Technology Protection" technology dubbed 'Terminator'. This yet-to-be-commercialized technology would allow the production of first generation crops that would not generate seeds in the second generation because the plants yield sterile seeds. The patent for this so-called "terminator" gene technology is owned by *Delta and Pine Land Company* and the United States Department of Agriculture. Delta and Pine Land was bought by Monsanto Company in August 2006. Similarly, the hypothetical Trait-specific Genetic Use Restriction Technology, also known as 'Traitor' or 'T-gut', requires application of a chemical to genetically modified crops to reactivate engineered traits. This technology is intended both to limit the spread of genetically engineered plants, and to require farmers to pay yearly to reactivate the genetically engineered traits of their crops. Traitor is under development by companies including Monsanto and AstraZeneca.

In addition to the commercial protection of proprietary technology in self-pollinating crops such as soybean (a generally contentious issue), another purpose of the terminator gene is to prevent the escape of genetically modified traits from cross-pollinating crops into wild-type species by sterilizing any resultant hybrids. The terminator gene technology created a backlash amongst those who felt the technology would prevent reuse of seed by farmers growing such terminator varieties in the developing world and was ostensibly a means to exercise patent claims. Use of the terminator technology would also prevent "volunteers", or crops that grow from unharvested seed, a major concern that arose during the Starlink debacle. There are technologies evolving which contain the transgene by biological means and still can provide fertile seeds using fertility restorer functions. Such methods are being developed by several EU research programmes, among them Transcontainer and Co-Extra.

PRODUCTS OF AGRICULTURAL BIOTECHNOLOGY

The regulatory approach to the safety evaluation of plants developed using rDNA technology has evolved in the best interests of research scientists, industry and the general public. The agricultural products of rDNA technology, such as GM foods and crops, may require approvals from up to three regulatory agencies; the US FDA, the USDA and the US EPA, depending upon the characteristics exhibited by the GM plant, its proposed use and introduced traits. The same standards of safety are applied to all products regardless of the technology used in their development.

The US FDA is responsible for ensuring the human safety of all new foods and food components, including products developed using rDNA technology, under the Federal Food, Drug, and Cosmetic Act (FFDCA). The USDA evaluates the potential of a GM plant to become a plant pest following its environmental introduction under the Federal Plant Pest Act (FPPA). The US EPA evaluates pesticides, including plant systems modified to express pesticides (*e.g.* insect-protected or virus resistance), under the Federal Insecticide, Fungicide, and Rodenticide Act (FIFRA). As a result, the expression of an insecticidal protein in a food crop would undergo review by the USDA, US EPA and US FDA; a GM food crop exhibiting a modified oil content would be evaluated by the USDA and US FDA; and a non-food horticultural plant developed using rDNA technology for any other purpose (*e.g.* flower colour) would be subject to review by the USDA alone.

In most instances, obtaining all necessary approvals for the commercialization of an agricultural crop developed using rDNA technology takes a decade or more. However, the exact amount of time required will depend on the need to confirm performance, to evaluate characteristics of the food, environmental effects, and to produce the required amount of seed before the product can be distributed and commercially grown by farmers. Up to five years of field trials (5-10 generations of plants) are required for the developer of a new plant variety to collect sufficient data to meet the reporting requirements of the USDA. An additional five months to two years may be required for the US FDA, USDA and/or US EPA to complete all necessary product consultations, reviews and approvals.

Approval for the first commercial planting of a GM food crop was not issued until 1995. Since 1995, more than 40 new agricultural crops developed using rDNA technology have received approval for commercial planting within the United States. In 1999, approximately 72 per cent of the total 39.9 million hectares (more than 98 million acres) of GM crops grown worldwide were planted in the United States. Herbicide-tolerant soybeans (54 per cent), Bt corn (19 per cent) and herbicide tolerant canola (9 per cent) accounted for approximately 82 per cent of the GM plants cultivated. With an increasing number of agricultural biotechnology products reaching later stages of commercial development, it is anticipated that the overall area planted with GM crops will continue to rise.

The regulatory approach to evaluating the human health and environmental safety of GM crops within the United States is best described as a science-based, case-by-case assessment of hazards and risks. This approach has provided the flexibility required to reduce the regulatory burden placed on products that have been determined to be of low risk or concern. All agencies involved in the regulation of plants developed using rDNA continue to implement and develop policies based on recommendations made within the Coordinated Framework. In the future, the US FDA, USDA and the US EPA will be dedicating additional resources towards communicating how GM food and food components are regulated within the United States, and how these regulations function to be protective of both human health and the environment.

FOOD AND FOOD COMPONENTS

The US FDA is responsible for ensuring the safety and wholesomeness of all food and food components, including the products of rDNA technology, under the FFDCA. The US FDA has the authority for the immediate removal of any product from the market that poses potential risk to public health or that is being sold without all necessary regulatory approvals. As a result, a legal burden is placed on developers and food manufacturers to ensure the commodities utilized and foods available to consumers are safe and in compliance with all legal requirements of the FFDCA. In order to understand the regulatory approach followed by the US FDA in the safety evaluation of GM crops, it is useful to consider food and food safety from a historical context.

People had been consuming foods derived from agricultural crops for many years prior to the existence of any food laws or regulations within the United States. Based on this experience, agricultural crops have been accepted as being safe for consumption as food, without additional testing to demonstrate their safety.

As long as the new crop variety has exhibited similar agronomic properties, and an appropriate taste and appearance, it has been considered safe to consume. As a result, most foods consumed today, in particular whole foods (*i.e.* fruits, grains and vegetables) and conventional foods, have not been subject to any kind of premarket review or approval by the US FDA. Nevertheless, food scientists have a good understanding that many of the commonly consumed agricultural crops contain natural toxicants (*e.g.* tomatine in tomatoes, solanine in potatoes, cucurbiticin in cucumber, psoralens in celery, etc.). As a result, new plant varieties may be subject to routine chemical analyses to ensure that none of these substances is present at potentially harmful levels. This type of general approach has been used in assessing the safety of thousands of new plant varieties that have been developed over a number of decades of crop breeding without compromising the safety of whole foods.

Role

Consistent with recommendations within the Coordinated Framework, the US FDA considered existing provisions of the FFDCA to be sufficient for the regulation of foods and food components developed using rDNA technology. It was concluded that the scientific and regulatory issues posed by the products of rDNA technology were not significantly different from those posed by conventional products. As a result, GM foods and food components have been subject to the same standards of safety as already exist for the regulation of other foods and food components under the FFDCA. In order to better communicate interpretations of existing provisions of the FFDCA as they relate to the safety evaluation of foods derived from new plant varieties, including the products of rDNA technology, the US FDA released a policy statement in 1992 entitled 'Statement of policy: foods derived from new plant varieties'.

The US FDA has considered the use of genetic modification (*i.e.* rDNA technology) in the development of new plant varieties to represent a continuum of conventional plant breeding practices (*e.g.* mutagenesis, hybridization, protoplast fusion, etc.), and as a result, the safety evaluation of all new plant varieties, not just those developed using rDNA technology, have been evaluated based on an objective analysis of the characteristics of a food or its components, and not on its method of production.

Definition and Scope of Bioengineered Foods

The recently proposed rule of the US FDA concerns 'bioengineered foods' which have been defined as 'foods derived from plant varieties that are developed using *in vitro* manipulations of DNA (generally referred to as rDNA technology)'. As a result, the proposed rule has a much narrower focus than the 1992 US FDA Statement of Policy.

The US FDA has explained the need for a change in emphasis based on their expectations that many of the new plant varieties exhibit a greater potential to 'contain substances that are significantly different from, or that are present in food at a significantly different level than before'. As a result, the substances present in foods and food components derived from new plants developed using rDNA technology are less likely to be considered GRAS, and as a result, will require pre-market approval from the US FDA.

Safety and Nutritional Evaluation

The primary objective of the safety and nutritional evaluation is to demonstrate that the food derived from a new plant variety is as safe or nutritious as foods already consumed as a part of the diet. For new plant varieties, including those developed using rDNA technology, a science-based approach is used to focus the evaluation on the demonstrated characteristics of the food or food component. The evaluation of a GM food or food component typically involves reviewing information or data on any newly introduced substances, the known levels of toxicants, as well as the nutritional composition of the plant following modification. Substances that raise safety concerns (*e.g.* toxicants, allergens) would be subject to more extensive evaluation, since both intended and unintended changes may affect the levels of toxicants and nutrients in a food following the modification.

Guidance for performing a safety and nutritional evaluation was provided in the 1992 US FDA Statement of Policy, in a series of flow charts and text that cover:

- The crop that has been modified;
- Source(s) of the introduced genetic material;
- New substances intentionally added to the food as a result of the genetic modification (*e.g.* proteins, but also fatty acids, and carbohydrates).

Documentation required to support the evaluation typically includes: the purpose or intended technical effect of the modification on the plant, together with a description of the various applications or uses, a molecular characterization of the modification including the identities, sources and functions of introduced genetic material; information on the expressed protein products encoded by introduced genes; information relating to the known or suspected allergenicity and toxicity of any expressed gene products; for foods known to cause allergy, information on whether the endogenous allergens have been altered by the genetic modification; information on the compositional and nutritional characteristics of the foods, including anti-nutrients; and in some instances, comparative results of feeding studies involving the foods derived from plants modified using rDNA technology and the non-modified counterpart.

In performing its evaluation, the US FDA is particularly interested in the identification of inherent toxicants, known or potential allergens, assessing the concentration and bioavailability of essential nutrients, the safety and nutritional value of any newly introduced proteins, and the identity, composition, and nutritional value of modified carbohydrates, fats and oils.

If additional questions of safety remain following this evaluation, further toxicological studies may need to be performed. It is recognized that absolute assurance of the safety of any food does not exist. As a result, the goal of the safety evaluation is to establish a reasonable certainty of no harm under anticipated conditions of consumption. With this in mind, experience with the existing food supply has provided the basis for evaluating the safety of new food or food components. Both the Food Advisory Committee and Committee for Veterinary Medicine have been extensively involved in the development of approaches for the safety and nutritional evaluation of foods and food components derived from new plant varieties, including those developed using rDNA technology.

MULTIPLIER EFFECT ON AGRICULTURAL PRODUCTION

Drought has multiplier effect on agricultural production during the subsequent year also, due to (i) non-availability of quality seeds for sowing of crops, (ii) inadequate draught power for carrying out agricultural operations as a result of either distress sale of cattle or loss of life, (iii) reduced use of fertilizers as the investment capacity of the farmers decline, (iv) non-availability of raw materials in agro-based industries, and (v) deforestation to meet the energy needs in domestic sector as agricultural waste may not be available in required quantity.

The Central Ministry of Agriculture (MOA) is responsible for implementation and formulation of national policies and programmes to achieve agricultural growth through optimum utilization of the land resources, water, soil, plant, fisheries, & livestock resources. Government of India implements the following agricultural related Schemes (whether Watershed based or Agro-climatic region based) in the country, which deal agricultural resources information for Planning and Development:-

- Agro-climatic Regional Planning (ACRP) Project
- Agro-Ecological Mapping Project of the National Bureau of Soil Survey & Land Use Planning (NBSS&LUP)
- All India Soil and Land Use Survey (AISLUS)
- Early Warning System of Agricultural Situation in India
- Forecasting of Agricultural output using Space, Agro-meteorology and Land based observations (FASAL) Project
- Land Records Computerisation Project
- National Agricultural Research Project (NARP)
- National Agricultural Technology Project (NATP) to strengthen research-extension-farmer (r-e-f) linkage
- National Watershed Development Programme for Rain-fed Areas (NWDPRA)
- Soil and Water Conservation Programmes
- Drought Prone Area Development programme
- Desert Development Programme
- National Wastelands Development programme
- Integrated Mission on Sustainable Development (IMSD) Programme.?

BIODYNAMIC AGRICULTURE

Biodynamic agriculture is a method of organic farming that treats farms as unified and individual organisms, emphasizing balancing the holistic development and interrelationship of the soil, plants, animals as a self-nourishing system without external inputs insofar as this is possible given the loss of nutrients due to the export of food.

Regarded by some as the first modern ecological farming system and one of the most sustainable, biodynamic farming has much in common with other organic approaches, such as emphasizing the use of manures and composts and excluding of the use of artificial chemicals on soil and plants. Methods unique to the biodynamic approach include the use of fermented herbal and mineral preparations as compost additives and field sprays and the use of an astronomical sowing and planting calendar. Biodynamics originated out of the work of Rudolf Steiner, the founder of the spiritual philosophy anthroposophy.

HISTORY

The development of biodynamic agriculture began in 1924 with a series of eight lectures on agriculture given by Rudolf Steiner at Schloss Koberwitz in what was then Silesia, Germany,. The course was held in response to a request by farmers who noticed degraded soil conditions and a deterioration in the health and quality of crops and livestock resulting from the use of chemical fertilizers.

An agricultural research group was subsequently formed to test the effects of biodynamic methods on the life and health of soil, plants and animals. Today biodynamics is practiced in more than 50 countries worldwide. Demeter International is the primary certification agency for farms and gardens using the methods.

Regional

In the United States, the Biodynamic Farming & Gardening Association was founded in 1938 as a New York state corporation. In Australia the first biodynamic preparations were made by Ernesto Genoni in Melbourne in 1927 and by Bob Williams in Sydney in 1939. Since the 1950s research work has continued at the Biodynamic Research Institute in Powelltown, near Melbourne Australia under the direction of Alexei Podolinsky. In 1989 Biodynamic Agriculture Australia was established, as a not for profit association. The University of Kassel has a dedicated Department of Biodynamic Agriculture.

BIODYNAMIC METHOD OF FARMING

Biodynamic agriculture conceives of the farm as an organism, a self-contained entity with its own individuality. "Emphasis is placed on the integration of crops and livestock, recycling of nutrients, maintenance of soil, and the health and well being of crops and animals; the farmer too is part of the whole." Cover crops, green manures and crop rotations are used extensively.

Biodynamic Preparations

Steiner prescribed nine different preparations to aid fertilization which are the cornerstone of biodynamic agriculture, and described how these were to be prepared. Steiner believed that these preparations transferred supernatural terrestrial and cosmic "forces" into the soil. The prepared substances are numbered 500 through 508, where the first two are used for preparing fields whereas the latter seven are used for making compost. A long term trial evaluating the biodynamic farming system in comparison with organic and conventional farming systems, found that preparations have influence on soil structure and micro-organisms enhancing soil fertility and increasing biodiversity.

Regarding compost development beyond accelerating the initial phase of composting, some positive effects have been noted:

- The field sprays contain substances that stimulate plant growth include cytokinins.
- Some improvement in nutrient content of compost.

Field Preparations

Field preparations, for stimulating humus formation:

- *500*: A humus mixture prepared by filling the horn of a cow with cow manure and burying it in the ground in the autumn. It is left to decompose during the winter and recovered for use the following spring.
- *501*: Crushed powdered quartz prepared by stuffing it into a horn of a cow and buried into the ground in spring and taken out in autumn. It can be mixed with 500 but usually prepared on its own. The mixture is sprayed under very low pressure over the crop during the wet season, in an attempt to prevent fungal diseases. It should be sprayed on an overcast day or early in the morning to prevent burning of the leaves.

Both 500 and 501 are used on fields by stirring about one teaspoon of the contents of a horn in 40–60 liters of water for an hour and whirling it in different directions every second minute. Although some biodynamic beliefs refer to buried quartz "fermenting".

Compost Preparations

Compost preparations, used for preparing compost, employ herbs which are frequently used in medicinal remedies:

- *502*: Yarrow blossoms are stuffed into urinary bladders from Red Deer placed in the sun during summer, buried in earth during winter and retrieved in the spring.
- *503*: Chamomile blossoms are stuffed into small intestines from cattle buried in humus-rich earth in the autumn and retrieved in the spring.
- *504*: Stinging nettle plants in full bloom are stuffed together underground surrounded on all sides by peat for a year.
- *505*: Oak bark is chopped in small pieces, placed inside the skull of a domesticated animal, surrounded by peat and buried in earth in a place where lots of rain water runs past.
- *506*: Dandelion flowers is stuffed into the peritoneum of cattle and buried in earth during winter and retrieved in the spring.

- *507*: Valerian flowers are extracted into water.
- *508*: Horsetail

One to three grams of each preparation is added to a dung heap by digging 50 cm deep holes with a distance of 2 meters from each other, except for the 507 preparation, which is stirred into 5 liters of water and sprayed over the entire compost surface. All preparations are thus used in homeopathic quantities. Each compost preparation is designed to guide a particular decomposition process in the composting mass. One study found that the oak bark preparation improved disease resistance in zucchini.

Astronomical Planting Calendar

The approach considers that there are astronomical influences on soil and plant development, specifying, what phase of the moon is most appropriate for planting, cultivating or harvesting various kinds of crops. This aspect of biodynamics has been termed "astrological" in nature.

Treatment of Pests and Weeds

Biodynamic agriculture sees the basis of pest and disease control arising from a strong healthy balanced farm organism. Where this is not yet achieved it uses techniques analogous to fertilization for pest control and weed control. Most of these techniques include using the ashes of a pest or weed that has been trapped or picked from the fields and burnt.

A biodynamic farmer perceives weeds and plant vulnerability to pests as a result of imbalances in the soil.

- Pests such as insects or field mice have more complex processes associated with them, depending on what pest is to be targeted. Field mice are to be countered by deploying ashes prepared from field mice skin when Venus is in the Scorpius constellation.
- Weeds are combated by collecting seeds from the weeds and burning them above a wooden flame that was kindled by the weeds. The ashes from the seeds are then spread on the fields, then lightly sprayed with the clear urine of a sterile cow, this is intended to block the influence from the full moon on the particular weed and make it infertile.

Seed Production

Biodynamic agriculture has focused on open pollination of seeds and the development of locally adapted varieties. The seed stock is not controlled by large, multinational seed companies.

AGRICULTURAL GROWTH AND PERFORMANCE

While the above analysis has provided a general view of the impact of economic reforms, this chaper examines agricultural growth and performance in the states of Bihar, Karnataka, Tamil Nadu and Punjab with their attendant policy implications.

The yields for various crops in these states differ greatly. While Tamil Nadu had the highest yield in rice, oil seeds and sugarcane, Punjab enjoyed the highest yields in wheat, coarse cereals, pulses and food grains. Karnataka on the other hand is seen to do well in cotton and Bihar performed quite well in pulses and coarse cereals. Further analysis and findings by Kalirajan and others (2001) show that Punjab had made remarkable achievements on the agricultural front while Bihar had remained stagnant in the last two decades, with Karnataka and Tamil Nadu showing moderate achievement. Clearly, differences in physical endowments, climatic conditions and institutional characteristics are some of the reasons for the varying productivity performance. Thus, having across the board economic reforms is likely to work less effectively than state-specific policy measures that enable each state's agricultural yields to reach their full potential. The comparative advantage of each state's agricultural production should be determined and with inter-state restrictions removed, total agricultural output would see a very significant increase.

For example, Karnataka with less favourable soil and water resources should be given incentives to concentrate on agro-processed products and corporate agriculture in horticulture, floriculture and animal husbandry, or to undertake watershed development to help with dry land agriculture. Many studies have indicated that with watershed areas, productivity growth has been mainly due to seed and fertilizer use. Thus, this state has to be given input subsidies for high yielding seed varieties but at the same time, the farmers need to be educated on the over use of chemical fertilizers.

With Bihar, agricultural performance is problematic on many fronts. First, although demographic pressure has increased and agricultural technology has improved, most of the uncultivated land is concentrated in southern Bihar, where irrigation facilities have not kept pace and the soil is of poor quality. Given the physiography of southern Bihar, wells are also unsuitable and thus the dominant mode of irrigation has been through tanks whose expansion and maintenance has been neglected. Second, the infrastructural facilities of Bihar have been lagging as seen by the infrastructure development. Due to infrastructural bottlenecks, availability of modern goods and services has not increased or their supply remains costly or unreliable.

Third, agriculture in Bihar is dominated by small and marginal farmers and the prevalence of mass poverty is largely related to the backwardness of agriculture. Fourth and importantly, the state agricultural policies in Bihar are in dire need of review. The semi-feudal production condition still exists in rural areas and the ineffective protection of tenancy rights has hindered agricultural growth. The slow pace of land consolidation reflects inadequate financial outlays and a shortage of manpower. Kalirajan and others (2001) note that marketing and extension services in Bihar are also rather weak compared to the other states.

Punjab on the other hand, was one of the few states which enjoyed the success of land reforms and the high priority of investment in rural infrastructure. Also, the irrigation base of the small and medium sized farms was comparable to that of large farms. In addition, the Punjab Agricultural University at Ludhiana contributed to the development of new seed varieties.

However, there are clear signs of a decline in crop yields since the 1990s and this has been associated with the increasing use of fertilizers and excessive water use which have increased the unit cost of production as a result of declining soil quality. Hence, care is needed when providing further input subsidies in fertilizer and water use. Another related fact is the steep increase in wages in Punjab and in the absence of productivity increases, the cost increase has affected the profitability of farmers. With Tamil Nadu, the main crop has been rice as this state is blessed with two monsoons. But from 1992-1997, there has been a steady decline in the areas irrigated by canals and an increase in well-irrigated areas while the use of tanks remains an unreliable source of irrigation.

However, major improvements in about 10 rice varieties released in the early 1990s can be expected to improve productivity growth in rice production although pests and diseases as well as imbalance in the use of fertilizers are major constraints. Thus Tamil Nadu could do with subsidies of pesticides and farmers should be educated on the more effective use of fertilizers to obtain high yields. Interestingly, the cropping pattern of late has shown increasing substitution of food crops by commercial crops but there is concern that the benefits will reach farmers only with the development of adequate infrastructure such as roads and markets. However, that Tamil Nadu has a higher index than the all India average of infrastructure.

Challenges of Globalization

It is important to realize that globalization poses many challenges to a developing country like India, which had relied on a state directed and regulated policy regime for more than four decades. In moving to a more open, market-based economy there are many transitional problems that the country has to manage. The Government must play a pro-active role in facilitating the globalization process so that the opportunity sets for the economic agents are widened and the adverse effects of globalization are minimized. The Indian Government must also prepare the necessary information base and develop its capacity to articulate India's concerns and policy trade-offs in the international forums for multilateral trade and environmental negotiations.

In addition, the Government should embark on an extensive programme to educate farmers on the need to meet the standards required in the export markets. In fact, India needs to seek technical assistance in creating the capacity for meeting such standards and to consider watershed developments for environmental considerations. Equally important is the need to disseminate information about possible export markets to farmers, so that market access is achieved at minimum cost. Given the requisite information about markets and profitability, the likelihood of farmers investing in post-harvest and processing technologies and storage and efficient transportation arrangements as well as developing supporting infrastructure is very high.

Although the brave and bold move by India to reduce the tariff rate for agricultural products from 113 per cent in 1990-1991 to 26 per cent in 1997-1998 deserves to be applauded, the question of whether India is ready to compete in world markets remains to be seen. The infant industry argument may still hold for India to shield itself from external competition but one can easily question the length of time that is required to that end. Also, a delay in opening up to foreign trade has the danger that local producers may become too complacent and never be ready for competition.

As India opens up externally, it is also expected to face vulnerability in the wider international price fluctuations and thus Acharya (1998) claims that a minimum price support scheme is important. These prices can also act as a signal to adopt modern inputs and invest in yield-raising infrastructure for increasing production. For instance, keeping basic staple food grains at reasonable prices would induce farmers to switch over to high value crops. However, during the 1990s, Kalirajan and others (2001) shows that procurement prices especially for rice and wheat have been increasing faster than the general price level. Such high prices along with guaranteed purchases by the Food Corporation of India have pushed up market prices. These higher prices are partly responsible for the large buffer stocks with the Food Corporation. If this trend continues, India's comparative advantage will be eroded.

With openness and high price instability, unstable export revenue can also be expected. One way of reducing such risk is for India to diversify her agricultural exports. For example, since 1990, even in commodities such as tea, coffee, cocoa and spices, where India is supposed to have a comparative advantage international prices have been unstable. Besides increasing the type of exports to obtain more export revenue, India should also seriously consider exporting more value added agricultural products through agro-processing such as processed vegetables, fruits, fish and meat products given that export or even local demand for basic agricultural products would decline as incomes rise. The move to higher value added activities within the agricultural sector also spells greater opportunities for industrialization and vice versa as borne by Kalirajan and Shand's (1997) findings of a bi-directional relationship between agriculture and industry for most Indian states. On the other hand, Sivakumar and others (1999) establish empirical evidence of high forward linkages of agriculture due to the presence of agro-industries while Satyasai and Viswanathan (1999) show the significance of the spillover effects to the industrial sector via the intensive use of purchased inputs in the agricultural sector.

The lack or slow pace of internal or domestic liberalization is also seen to hinder the possible gains from external or trade liberalization. For example, although central zoning restrictions have been abolished, state government restrictions on inter-state and even inter-district restrictions on marketing and movement of goods still exist in many cases. This interferes with the benefits from crop specialization and economies of scale arising from comparative advantage. The land market is another example of distortion whereby land ceilings exist preventing the operation of large-sized farms.

This has led to the emergence of a large number of small economically unviable land holdings. The easy leasing of land should be permitted with assurance of resumption. Yet another problem lies with the insufficiency of credit to agriculture. From 1995-1996, the Rural Infrastructure Development Fund was set up to allocate funds for the completion of projects and the government has committed itself to strengthening the cooperative credit structure through substantial refinancing and restructuring of the Regional Rural Banks. However, as mentioned earlier, due to varying institutional factors in the Indian states, these domestic reforms can be expected to yield quite different results.

Although India missed the opportunity to open up two decades ago, its attempts to do so now must be regarded as better late than never. Others such as Desai (1999) observe that, "the logic of the global economy as well as India's interests dictate that India become proactive in its liberalization policies. India must liberalize not because it has no choice but because it is the best choice".

His lament that India has adopted a 'victim mentality' when it really needs to adopt a 'winner mentality' has become less of a concern as over time, India has shown commitment to stay on the bandwagon of globalization. Having realized that globalization is a necessary but not a sufficient condition for high growth production, India has undertaken economic reforms, both internal and external. However, it must be ensured that these reforms are synchronized so that the pace of both reforms is set right in order to work hand in hand to promote agricultural productivity growth.

Thus, training the farmers and educating them appropriately to change their mindset and reorienting them to take up new activities or adopt foreign technology is of utmost importance. In this context, it is necessary to involve non-governmental organizations in training and mobilizing the rural poor to face the challenge of liberalization.

Also, with domestic economic reforms, more care needs to be exercised to draw up state-specific liberalization measures to maximize their benefits. Lastly, in the implementation of these reforms for successful globalization, one crucial element, not entirely within control is the need for good governance and stability in the political and economic environment. Political leaders who are the ultimate decision makers in these matters need to examine their own role dispassionately.

It is quite apparent that at this relatively early stage, there is little observable evidence of gains to India's agricultural performance after opening up. However, there could easily be benefits that have not yet surfaced, or are yet to be identified and perhaps too difficult or intangible to measure.

Whatever the case, it is highly likely that it is too soon to assess the full impact of globalization and economic reforms. Furthermore, the process of liberalization has been gradual and remains incomplete. For example, the complete removal of quantitative restrictions after March 2001 will have provided an opportunity for Indian farmers to tap world markets and, if they are successful, results should start to become evident soon. Export promotion via the development of export and trading houses as well as effective liberalizing export promotion zone schemes for agriculture are fairly recent measures and only time will tell as to how effective these measures are. Other possibilities such as agro-industry parks for promoting exports are also in the pipeline.

In conclusion, India has successfully set sail on the waters of globalization and economic reforms and even in the wake of economic and political instability, she has to carefully steer her course in order to reap the benefits of increased productivity growth in the agricultural sector.

TECHNOLOGICAL CHANGE IN AGRICULTURE

Technological change in agriculture has been the foundation of economic growth and development. By raising yields and continually increasing the number of people who can be fed by the output of one worker in agriculture, change in agricultural technology has permitted an increasing proportion of labour and other productive resources to be devoted to non-agricultural production. Plant and animal breeding and chemical and mechanical technology plus improved husbandry have caused continual structural change in farming. These technological stimuli have arisen as a series of overlapping waves to create a process of technological change, the momentum of which looks likely to be maintained by intensified adoption of information and biotechnologies.

As a result of the cumulative processes to date, farms have become larger and more specialized, and farmers are more highly trained and have substituted machines for animal and human power; farming in the 'West' has become one of the most capital-intensive of industries. At the same time agriculture has become more dependent upon industry, and even more industrialized-the intensive production of eggs, broiler chickens, horticultural products and pigs has many characteristics which belong to industrial production rather than to traditional farming. Goodman *et al.* talk of 'the industrial appropriation of the rural processes'. Processes of marketing farm products have passed off the farm to be performed by sophisticated distribution and retailing sectors which increasingly dictate details of production to the farm sector, while at the other end of the chain an increasing proportion of inputs is supplied by the chemical, pharmaceutical and engineering industries.

The process of 'industrializing' agriculture has been transferred to the less-developed countries (LDCs) and is epitomized by the so-called 'Green Revolution'. This has entailed expansion of irrigation systems using modern pumps, engineered dams and canals, plus the increased use of inorganic fertilizer, insecticides, herbicides, fungicides and power machinery; although at its heart has been the 'old' biotechnology of breeding new cereal varieties which give high yields when supplied with chemical inputs. Because much of the impetus for this 'revolution' has been western technology and farming know-how, particularly that of the USA, there are those who are critical of what they see as the LDCs' increasing technological dependence upon the West, its industries and research institutions. What is undeniably the case is that technological advance has been faster, and public and private expenditure on research in western agriculture greater than in LDCs taken as a whole.

Thus exportable surpluses from the West have increased with generous public support, while many LDCs have had to resort increasingly to food imports to meet their needs. While at this stage it is difficult to foresee exactly when, and on what commercial scale, the new agricultural biotechnology 'revolution' which is now brewing will have its effect, most commentators agree that it will intensify the industrialization of agriculture, and that it will increase the technological dependence of most LDCs upon western firms.

It may cause considerable disruption to the economies of some countries and will cause farming operations to diverge increasingly from the pattern associated with the country yeoman. Rural life in many more remote farming areas will be further threatened by the continued deterioration in the economics of ruminant livestock farming on upland and low-productivity pastures. It is these redistributive and structural effects that are the central focus here rather than any attempt to estimate the scale and speed of uptake of specific biotechnology inputs and processes.

Although prospective biotechnological change in agriculture is not different in many fundamental respects from other processes of technological change, in that it will operate through the provision of new or modified inputs (seeds, hormones, etc.) and the creation of new markets in industry, it does raise new issues.

Are there dangers from releasing life forms which are engineered using new methods which are quite different from those associated with varieties developed by traditional plant-and animal-breeding methods? Even if the dangers can be objectively assessed and turn out to be minimal, will what may be perceived as man-made and therefore 'unnatural' products be acceptable to society as a whole?

AGRICULTURAL TECHNOLOGIES AS EXPERIMENTS

Farmers have continuously, and quite consciously, experimented in their fields with crops. Most observers have overlooked this process, emphasising instead innovations that have originated in laboratories and field stations of research organisations and corporations established specifically for that purpose. Yet, 'probably, the total genetic change achieved by farmers over the millenia was far greater than that achieved by the last hundred or two years of more systematic science-based effort'. In suggesting that developing country farmers were keen to utilise new technology if only it were available to them, a view deemed radical at a time when developing country farmers were visualised as both backward and conservative, Theodore Schultz implied that the technology they were employing was not changing. Similarly, Jackson and Ford-Lloyd in suggesting that artificial selection by humans was 'probably of largely unconscious nature', seem to devalue the contribution of early farmers in developing crop species.

Probably, Cox and Atkins are right to suggest some traits were selected for consciously, others by default. Nevertheless, the fact that seed selection has been practised all over the globe for millennia supports the view that one does not need a rigorous understanding of genetics to recognise value in such practices, and as Biggs and Clay point out, farmers work in continually changing environments. Without experimenting with crop varieties, most of the plants which today we recognise as crops would never have evolved to their current form.

DIVERSITY IN DECLINE

Since the mid-nineteenth century, an increasing amount of research into plant breeding has taken place in organisations set up partly for that purpose. As plant breeding became more successful, producing cultivars matching the needs of an increasingly market oriented agriculture (expansion of which was assisted by breeding), so there began a decline in the genetic diversity of crops in the field. Holden *et al.* suggest that crop genetic diversity probably peaked around the middle of the nineteenth century. The double bind which faces those who have sought to replace landraces, or diverse crop mixtures, with uniform 'improved varieties' is that the genetic diversity on which their activity depends may be lost as a consequence of their very success. This is one way in which 'genetic erosion' occurs.

The value of landraces for the future of agriculture was recognised by two German scientists, von Proskowetz and Schindler, at the end of the nineteenth century. Nikolai Vavilov then laid the basis for work on centres of origin and diversity of crop plants, suggesting that the genetic diversity of a given crop coincided with its centre of origin. An exact correspondence has been refuted, and today Vavilov's localised centres are discussed in terms of centres and non-centres which cover significant geographical areas.

However, the idea that some areas of the world exhibit great diversity for particular crops continues to be widely accepted, and is of crucial importance to breeders today as genetic erosion gathers pace, even in these areas. Loss of diversity has taken place both between and within species, and as a result, it can be argued, in the range of techniques employed in agriculture: 'the mosaic appears to be disappearing'. As Wilkes notes, a shift from growing crops for consumption to growing crops for sale has resulted in:

the dropping of many legume crops (20–35 per cent protein) which have been classic protein sources in the human diet for more fertilizer/water-yield responsive cereals (7–14 per cent protein). The current situation is a rapid reversal of eight millennia of *in situ* evolution where the cultivators ate the crops they cultivated.

Today, of more than 200 000 plant species currently known to scientists, only 30 are used to supply humans with the bulk of their food. If one was to compile a list of the top five crops in every country, only 130 species would be named. It should come as no surprise that there are many 'unknown' crops ideally suited for production and consumption purposes in the regions where they have been selected. Of those crops that are widely grown, the extent of diversity exhibited within the species is also narrowing. This entails risks in terms of the food supply's vulnerability to catastrophic losses, though as with many other ecological problems, quantitative estimates of such risks are subject to great uncertainty.

Even if the probability of such epidemics occurring were low (which we do not know), they may have severe consequences: 'Loss of genetic diversity – silent, rapid, inexorable – is leading us to a rendezvous with extinction – to the doorstep of hunger on a scale we refuse to imagine'. Given the prevailing uncertainty, it is equally valid to depict optimists as complacent as it is to brand the authors of the above words as scaremongers. Such is the state of knowledge about the system which feeds many of us.

Estimates as to the extent of genetic erosion tend not to take on a reliable quantifiable form, partly because by the time the problem began to be addressed, much diversity had already disappeared. Although the loss of varieties is nothing new, it is the pace of recent changes that is without precedent:

Since World War II, for example, virtually all of the local wheat varieties in Greece, Italy and Cyprus have been abandoned and most of the indigenous sorghum (*Sorghum bicolor*) races of South Africa disappeared after the release of high-yielding Texas hybrids.

But if the fact that genetic erosion is occurring is beyond doubt, the manner in which it occurs is disputed. Even those who emphasise the contributions made by traditional agriculture through crop selection have tended to fall into the trap of believing that the use of modern varieties always implies loss of traditional varieties. That they do so is surprising given that many seek to highlight the wisdom of indigenous practices. Why would such wise cultivators simply discard varieties they have so carefully selected and take a leap into the unknown? The work of Brush and others suggests that in some cases they do not:

The dismissal of *in situ* conservation is based on an extrapolation from temperate or from atypical regions of the tropics. This reasoning rests on conventional ideas about modernisation, which envision a linear replacement of earlier stages of agricultural development by more advanced ones... we should re-examine our assumptions about the loss of diversity, thus opening the plausibility of on-site conservation.

How new varieties are introduced and used in agro-ecosystems varies from place to place and from farmer to farmer. It is not impossible for modern varieties to augment diversity, or lead to the production of novel populations. Where this occurs, it reflects *diversity's value in use*. Even where replacement does occur, the value of discarded varieties may subsequently become apparent.

The Risks of Genetic Uniformity

Why is genetic erosion a problem? In traditional agricultural systems, epidemics are rare, unnoticed, undocumented, or all three. This is partly due to the genetic diversity, and hence, flexibility retained in the agricultural system. Genetically uniform systems lack this flexibility. The more uniform the crop over space and time, and the less frequent are environmental conditions deleterious to particular pests and pathogens (they are most infrequent in the tropics), the more likely it becomes that an epidemic will occur, and occur over a large area. Modern agricultural systems are characterised by their uniformity. Ecologically, they are artificial in the extreme, and the level of inter- and intra-specific diversity is, in terms of the system's ecological vulnerability, worryingly low. The possibility that pest epidemics will emerge increases due to the strong selection pressure exerted in favour of pests which can overcome any resistance that the crop may have had. In 1970, nearly a quarter of the US maize crop was wiped out by southern corn blight owing to the uniformity of the cytoplasmic DNA of hybrids making each equally susceptible to the fungus *Helminthosporium maydis*.

A subsequent National Academy of Sciences investigation concluded that, not just in maize, United States agriculture was, 'impressively uniform genetically and impressively vulnerable'. Among those alarmed by developments was Jack Harlan, whose father Harry had warned as long ago as 1936 that: When new barleys replace those grown by the farmers of Ethiopia or Tibet, the world will have lost something irreplaceable. When that day comes then our collections, constituting as they do but a small fraction of the world's barleys, will assume an importance now hard to visualise.

The significance of Harry Harlan's comments lay in his awareness that genetic variation is what breeders require to develop new varieties, and lost varieties may contain alleles of genes whose importance cannot be estimated in advance. Well before Harlan made these observations, the same uniformity had led to:

- The potato famine in Ireland;
- The French wine industry being struck by powdery mildew in 1848;
- The destruction of the coffee industry in Ceylon (then the world's largest producer) in the late nineteenth century; and
- Two wheatless days a week in the US in 1917 due to red rust.

Since the corn blight episode, uniformity has led, amongst others, to problems with rust in the UK wheat crop in 1972 owing to widespread planting of two wheat varieties (supposedly resistant to rust); orange wheat blossom midge in the UK in 1993; rice blast in South Korea in 1980 when 75 per cent of rice lands were sown with modern varieties of similar pedigree (forcing the country to import rice; Russian wheat aphid in a number of countries including the US and growing problems in 1990s US and the UK with the fungus responsible for the Irish potato famine, *Phytophthora infestans*. Of all these episodes, only the case of potatoes in Ireland can be said to have created a famine. Yet the potential exists for widespread crop losses, and in a world where global food stocks are shrinking, the likelihood of such events occurring in the future may be increasing.

The standard approach to dealing with vulnerability associated with genetic uniformity has been to breed for pathotype specific (vertical) resistance to prevent losses due to a particular ecotype of a pathogen. For the last fifty years or so, this 'disadaptive breeding' has been the approach of choice for formal agricultural research. Yet vertical resistance exerts the strongest selection pressure in favour of 'matching' pathotypes, limiting its useful life. Because the search for vertical resistance makes the degree of horizontal resistance difficult to detect, the level of horizontal resistance tends to be lower than is desirable, resulting in a boom-bust cycle. This is the so-called vertifolia effect. Breeders partake in what has become known as a varietal relay-race in their efforts to pre-empt the devastation of crops by pests and pathogens.

In situation's where vertical resistance is matched, or where vertical resistance cannot be found (because it may not exist), the use of pesticides is unavoidable. The use of pesticides, in turn, has led to development of new ecotypes of old pests and pathogens which are resistant to chemical attack, as well as the emergence of new pests and pathogens that thrive in the absence of their natural. When spraying to kill some pests, particularly those whose vectors move easily between fields, farmers create positive benefits for themselves, and a positive externality for other farmers in that they help to contain populations. Yet they may also incur costs to themselves in terms of health impacts, and lost productivity as a result, as well as two types of negative externality. The first relates to environmental pollution.

The second relates to the fact that pest resistance builds up at a rate determined in part by the volume of the active ingredient used. All farmers are adversely affected in the medium term through yield loss owing to existing resistance being overcome. Uniformity is, therefore, a problem with at least two related dimensions; vulnerability, leading at best to market instability and loss of earnings, and at worst, to widespread hunger; and associated environmental pollution.

Alternative strategies to vertical resistance include:

- Breeding for pathotype non-specific (horizontal) resistance – horizontal resistance, though more difficult for plant breeders to achieve, is rather more durable. Those favouring vertical and horizontal resistance can be said to be adopting, respectively, Mendelian or biometrical approaches to the problem. The problems associated with vertical resistance are increasingly well recognised and a shift is occurring towards horizontal resistance strategies. However, horizontal resistance strategies are not without problems of their own, one of which is that it is not known how effective horizontal resistance strategies will be in the absence of agrochemical use.
- Diversifying the varietal make up of a crop in any given region – by planting varieties incorporating different vertical resistance genes in neighbouring fields, the possibilities of disease epidemics occurring are reduced. The method is most effective where an early variety is the source of reproduction for a pathogen of a late variety. Similarly, where a parasite is known to reside in a particular region in a given season and then spread in subsequent seasons, growing varieties with different resistance genes along its path may slow down, or prevent, development of related disease. In a slightly similar vein, Wilkes talks of the use of 'genetic firebreaks', bands of diverse germplasm employed to limit the progress of any epidemic.
- Varietal rotation. Varieties incorporating different resistance mechanisms are grown once every few years and rotated with others.
- Incorporating, in a given field, diversity within varieties, between varieties, or between species. At one end of this spectrum are multi lines and mixtures, in which different vertical resistance genes are either backcrossed into a single genetic background or used in distinct varieties. At the other end are the diverse systems in home gardens, and of permaculture and agroecology, where both inter- and intra-specific diversity are encouraged.

One can appreciate that these different recommendations require different responses from breeders, farmers and policy-makers. The methods vary in terms of their genetic flexibility. This may be a consideration that increases in importance if climates change quickly (in historical terms), causing 'environment erosion' of any resistance (vertical or horizontal) present in monocultures. Furthermore, use of diversity within the agroecosystem may offer the best prospects for eliminating the use of synthetic pesticides, though some advocates of horizontal resistance do so on the basis of its performance in this respect.

Risks associated with a narrow genetic base are reduced by widespread agrochemical use (with attendant consequences for the environment), and by substantial investments by public and, increasingly, private organisations whose aim it is to keep hold of the batton in the varietal relay-race, either through developing new chemicals, or new varieties.

This places a considerable burden on these organisations to continually come up with the goods in the face of unexpected and unannounced attacks from pests and pathogens, as well as (possibly) changing climate. As pathogens overcome vertical resistance conferred by individual genes or become immune to chemical attack, and as genetic erosion continues in crop plants, weedy relatives of crops are assuming greater importance. Harlan has been moved to comment that 'wild relatives stand between man and starvation.'

The suggestion that genetic erosion is a problem does not inevitably lead to the recommendation of a re-direction of research in favour of practices that maintain, enhance, or encourage diversity *in situ*. Many advocate a more comprehensive, better described, and better co-ordinated system of *ex situ* conservation in gene-banks. Those who support this approach have confidence in *ex situ* conservation to provide adequate back-up for a breeding system which they implicitly trust to overcome any future problems.

This position is increasingly being undermined by well-documented limitations of gene-banks, leading to growing recognition that *ex situ* storage of germ-plasm is inadequate of itself to ensure a secure genetic foundation of agriculture into the future.

This has led some to advocate acknowledgement, and some form of reward system, for the role played by various farmers and communities around the world in maintaining crop genetic diversity *in situ*. Within this school of thought, there is a difference in emphasis between those who advocate the conservation of genetic diversity *in situ* for use in future breeding programmes, and those who recognise diversity as a desirable property of agricultural systems in itself.

The former view lends itself open to criticism on grounds that it can be taken to imply that *in situ* diversity would inevitably disappear unless prevented from doing so. But the latter view, exemplified by the work of Brush and others, immediately suggests a need for reappraisal of the objectives of formal agricultural research. If diversity is of value in use, why do researchers not base their investigations more firmly on the incorporation of diversity at the inter- and intra-specific level? If farmers are capable of experimenting with novel germplasm, why do *ex situ* gene-banks fail to promote an active exchange of their resources with farmers?

External-Input Agriculture

It is often stated, indeed, it is frequently taken as axiomatic that modern high-external-input agriculture (HEIA), unashamedly uniform genetically, constitutes the only possible way to feed the worlds large and growing human population.

There are many who believe that to retain genetic diversity, in whatever way, within the agricultural system necessarily imply an economic loss, or at least, a loss in production. Supporters of this view point to the achievements of the Green Revolution in Asia and parts of Latin America. Lipton and Longhurst write: 'history records no increase in food production that was remotely comparable in scale, speed, spread and duration'. This most probably is the case, yet the same could probably be said for the accompanying narrowing of the genetic base, and indeed, the 'scale', 'speed' and 'spread' factors each have negative ecological manifestations.

Many who discuss the Green Revolution tend to limit their number of counterfactuals to one, a continuation of the *status quo*. Is it to be understood that agricultural research could only work effectively if it concentrated on monocultures? Could production increases not have been achieved in any other way? Is it impossible to construct plausible counterfactuals paths for change which left agriculture less vulnerable in terms of its genetic inflexibility, less dependent on crop protection chemicals, and yet with the potential (in the sense that it meets the sufficiency criterion) to feed the world?

It seems almost heretical to write these words, but even on grounds of productivity alone, it may be true to say that (and this may require qualification with respect to ecological regions) agroecosystems incorporating greater species diversity are the best performers in terms of output per unit area and/or economic performance. At the same time, they can offer a number of environmental benefits relative to HEIA (depending on the system) which may include energy conservation, reduced agrochemical use, soil conservation and resource use efficiency. Hence, there are three important, if for the most part obvious points to be made with respect to the position that HEIA is the only way to feed the world:

- Modern agriculture does not have, as one of its aims, feeding the world. It is principally geared to making profits;
- Farmers have a number of reasons for adopting one or other technology. Concern for global food supplies and their distribution does not weigh significantly in such considerations; and
- Nobody has, and nobody could, prove the truth of what is being stated. The choice with regard to approaches in agricultural research is far wider than is often understood, and suggesting that one or other is 'the best' (let alone 'the only') is likely to prove problematic, especially when so few of these approaches have been researched thoroughly.

There is reason to believe that more will be known about any given way of doing agriculture as more research is conducted upon it, raising the possibility for what we might term increasing returns to advocacy. In other words, if through gaining an early lead in the command of research resources, a particular approach improves its position relative to alternative approaches, it is more likely to command research funds in the future. Whilst Plucknett criticises those who argue the need for research on marginal lands (as opposed to the most productive lands), Lipton, mindful of the over-simplifications implied by researchers in dividing land into advanced or backward areas, asks the question:

Are the 'lead areas', often no longer the most promising (especially after much research with a long period of diminishing returns), but rather the areas that have been favoured by AR [agricultural research] in the past, and continue to be favoured because that is what scientists know how to do best?

Two questions need to be answered; why has one particular way of doing agriculture come to be so widely embraced the world over; and what are, what has happened to, and what might be the alternatives?

Comparing Farming Systems

In order to understand why it is at least somewhat odd that a particular way of doing agriculture has come to be accepted so broadly, it is worth considering some of the factors that might impinge upon the decision to farm in whichever way. The problems encountered when trying to compare farming systems are many and varied. They relate to:

- *Location specificity* – the only satisfactory way to compare two farming systems is not available to us. We cannot compare different systems at the same time on the same piece of land. Because the quality of land will be affected by past activities, a comparison conducted sequentially on the same piece of land will introduce errors. If the comparison is not to take on the form of an experimental test (and it is generally appreciated that research station performance differs widely from that in the field) one faces the problem of finding working farms which are not only similar in agroecological characteristics.
- *Motivations for farming* – farming systems are part of wider social systems and are embedded within a social context. Not all farmers have the same motivations, particularly if farming is not the primary source of income. Even where income maximisation processes are of importance, they may still take on different forms. Many farmers will seek to optimise performance with respect to multiple criteria. Reintjes *et al.* consider productivity, security, continuity and identity to be important objectives, any combination of which could constitute the guiding rationale for farming. Problems that might arise from a desire to control (if this were possible) for variation in

motivations include the possibility that choice of farming technique may be, indeed almost certainly is, endogenous to these motivations.

- *Managerial capabilities* – that one farm 'does better' than another in a given comparison could simply be regarded as proof of the superiority of one farmer's management skills. But it is difficult to dissociate the issue of management from that of the farmer's objectives, and indeed, those of the researcher. One might also ask whether the adoption of one or other farming system should itself be considered an issue in measuring management capabilities. Furthermore, the adoption of one or other system might *require* of farmers that their management capability improves. Perversely, this might raise questions as to the wisdom of farmers deciding to practice more managerially demanding styles of agriculture.

Different types of farming systems probably require completely different types of managerial skill, if that is the correct phrase. These considerations serve to caution us against treating technologies and techniques as isolated parts of what is actually a complex whole, in which the ability and resources available to manage the farm are often conditioned by the characteristics of the farm household. How the household functions as a unit, and how it interacts with the wider community, will affect the types of decision taken regarding farm management. *Duration of comparison* – farming has always been a risky enterprise, and although in different farming systems the principal source of those risks (and possibly where the burden of risks falls) is altered, it continues to be so. Issues of seasonality, year-to-year variation, and secular trends or even breakdowns in one or other variable in the system are of crucial importance.

Most of these can only be revealed through a comparison which takes the longer-term into account. On the other hand, such studies of a longer-term nature as have been done have taken the form of experiments which, due to the degree of control that is desired, tend to ossify the production systems under investigation. This prevents any evolution in the system that might be desirable or necessary over time. Given that, for many systems, design and its improvement are of supreme importance; such comparisons lose relevance over the longer term. *Differences in systems* – farming systems may be so different that they simply are not comparable in any meaningful sense of the word.

The sort of alternative systems considered above incorporate greater diversity on the farm, leading almost inevitably to a more diverse output than is characteristic of modern, highly-specialised farms. How can one really compare the output of such different systems, particularly when one also considers the many other factors mentioned above which mitigate against any easy comparison? Research methodologies which use the manipulation of a restricted number of variables as the basis for comparing organic and high-input agriculture have probably missed the point of, for example, organic agriculture altogether. They suggest an inability to appreciate that in such agriculture, what is sought is system integration through the interaction of components, and this is a goal rather than an instrumental means to a particular end. *The comparison to be made* – to the extent that a focus on one or more indicators inevitably fails to consider all the dimensions of the farming system, any comparison will be less than complete, and therefore prone to biased conclusions. If such comparisons are attempted, it seems impossible to dissociate the comparison from the motivations of researchers. Additionally, if a comparison is made on the basis of chosen indicators, for the comparison to be a fair one, the farmers being compared should indicate an equal weighting of these indicators in their subjective assessment of farm performance.

Clearly, comparison of alternative farming systems is extremely difficult. Yet for all these difficulties, the homogenising tendencies of the expanding agro-food system are difficult to deny, still more so to resist. Why do so many farmers apparently chose independently to farm in a broadly similar way the world over? Have the options open to them been inadequately represented? To state that alternative options have received inadequate attention from formal agricultural research is partly to answer the question, 'why are these options called alternatives?' Consideration of this issue is important in the light of the genetic vulnerability inherent in the prevailing practice of agriculture, and the environmental problems this creates.

Schools of Alternative Agriculture

As applied to agriculture, my use of the term 'alternative' is culturally biased. In some regions, modern high-input agriculture is simply unfeasible and 'alternative' agriculture is all that is available. Furthermore, several alternative movements have referred to the validity of indigenous technologies which have, it goes without saying, pre-dated such movements by some considerable time. Harwood comments that the roots of alternative agriculture are to be found in the discourse evolving at the turn of the twentieth century juxtaposing holism and reductionism.

Alternative agriculture's development parallelled that of modern agriculture, leaning on the insights of the growing number of soil scientists, and influenced by Darwin, who had a great interest in the biological attributes of soil. In the early twentieth century, several books were written which alluded to the complex interrelationships between components of the farming system that had been ignored or downplayed by agricultural scientists. The work of King and Howard on highly productive systems of agriculture in Asia helped reinforce this message. However, alternative movements seem to have been severely affected by the Second World War. For practitioners of biodynamic agriculture, based on Steiner's anthroposophy, the impact came through ideological channels, as the National Socialists banned their activities. Harwood cites a number of works which appeared in the 1940s and 1950s, a common theme of which was 'the increasing environmental harm and resource degradation brought about by 'modern' farming methods. Yet, for Harwood, the 1960s were a transition period of 'narrow focus', which saw the rise of modern agriculture and the parallel demise of alternatives in agricultural scientific discourse. Harwood's view is corroborated by the experience of different nations in the post World War Two era, during which the state has played a crucial role in the promotion of modern agricultural techniques.

Most interesting to note in this context is the way in which the prosaic critique by Rachel Carson was greeted by Harwood and his colleagues:

There was little or no debate during those years, in the biological sciences at least, on development direction. The success of current technologies was so overwhelming that it stifled serious debate of alternatives. The alternative farming 'schools' were practically non-existent and certainly in disrepute. I remember clearly graduate school discussions in 1964 about the 'crackpot Rachel Carson and her whistling in the wind' against the great benefits of DDT. In looking back on those heady years, I wonder about our arrogance and narrowness of vision.

ORGANIC FARMING AND AGRICULTURAL PRODUCTION

Strictly speaking, all agricultural production is 'organic' in that every product of agriculture is composed of organic substances. Among the earliest proponents of organic farming were two Britons, Sir Albert Howard and Lady Eve Balfour. Before World War II Howard advocated adoption of production systems that used no synthetic fertilisers or pesticides. For more than 30 years, on her research farm, Lady Balfour conducted experiments in which conventional British farming systems were compared with 'organic' systems free from agro-chemicals. Rodale, the American high priest of organic farming, may have received his inspiration from the published work of Howard and Lady Balfour. Food products from organic farming represent a relatively small but slowly growing proportion of foods purchased in Europe and North America. Across the EU, the agricultural area devoted to organic farming has increased from 0.5 per cent in 1993 to *ca* 3 per cent in the year 2000. The value of organic foods sold in the United States is estimated at $10 billion, across the EU at 8 bn Euro. The total value of retailed processed foods in the United States is over $500 billion, in the EU more than 350 billion Euro.

The *Codex Alimentarius* guidelines for organic farming specify the following objectives:

- to enhance biological diversity;
- to increase biological activity and maintain fertility in soils;
- to restore soil nutrients by recycling organic wastes;
- to rely on local renewable rather than imported manufactured resources;
- to promote healthy use of soil, water and air;

- to minimise pollution from agricultural practices.

Despite the interest of environmentally conscious consumers, rational development of organic farming has been hampered by confusion among concepts and definitions, comparable to diverse interpretations about 'sustainable agriculture'. A comprehensive review and state of progress among EU countries has been published. Another European publication estimates sales of organically produced foods among EU members at about 1 per cent of total food sales. It also comments on the higher prices for organic foods charged in supermarkets. In France price differentials between organic and non-organic produce range from 10 to 40 per cent for dairy products, close to 50 per cent for fresh fruits and vegetables [EEC 2002]. In an 89-page document the European Council presents regulations that specify permitted production, labelling and inspection practices for organic foods produced and sold in EU markets.

The general principles of organic farming are unquestionably admirable: harmonisation of agriculture with nature, protection of ecologies and environments, conservation of soil fertility and water resources, recycling of wastes, humane husbandry of livestock. At present organic foods are purchased by a small proportion of dedicated consumers who can afford the higher prices. Prices may gradually fall if and when demand and production increase. Whether or not strict adherence to EU regulations, total prohibition of all synthetic inputs and transgenic crops can ensure sustainable food security for all the poorest nations and communities is open to question.

CHEMICAL FERTILISERS

While those who claim to practise or encourage organic farming vary in specific recommended practices, most urge systems more environmentally benign, relatively free from synthetic pesticides, growth hormones and antibiotics. Whether chemical fertilisers should be permitted is not consistent among organic farming advocates. Fertilisers provide nitrogen, phosphorus, potassium and other elements to soils naturally deficient, or to replace soil nutrients depleted by growing crops. Chemical fertilisation began with importation of guano from sea birds in Peru and nitrates from deposits in Chile.

In addition to the main constituents – carbon, hydrogen and oxygen – 13 elements are known as essential plant nutrients largely taken up through root systems. Most essential nutrients had their origins in the earth's crust from which soils were formed by weathering when nutrients were released, dissolved in soil moisture, then absorbed by the root systems of wild and cultivated plants. The chemical composition of the earth's crust is highly diverse across the Planet; soils in different locations vary significantly in plant nutrient content (ICSU-CASAFA 1991).

Chemical fertilisers are compounded to provide nutrients in which soils are naturally deficient or depleted by cultivated crops. Nitrogen, essential for all plant growth, originates in the 75,000 tonnes of the gas which float above every hectare of the earth's surface. Elemental atmospheric nitrogen cannot be taken up by plants directly but, by complex biochemical reactions and various organisms, must be converted to ammonia or soluble nitrates. Microorganisms known as Rhizobia live in root nodules of leguminous plants where they convert atmospheric nitrogen into ammonia to fertilise the host plant. In industrial processes, elemental nitrogen is 'fixed' to produce chemical fertilisers in a form that release nitrogen from soil as and when needed by cultivated crops. Substantial energy is absorbed in conversions of nitrogen into ammonia. In a microbial conversion, energy liberated from fourteen moles of adenine triphosphate [ATP] is required to produce one mole of ammonia.

In a Haber-Bosch industrial reactor, pressures of *ca* 20 million Newtons/ m^2 at 500°C energise the combination of nitrogen with hydrogen. Some nitrogen-fixing bacteria obtain their energy from organic materials in soils; in legume root nodules Rhizobia derive their energy from the host plant in return for which they supply the plant with accessible nitrogen. As Rhizobia fix nitrogen in legume root nodules, therefore, some organic farmers rotate or intercrop legumes with cereals that lack the physiological and biochemical mechanisms to fix nitrogen. There is hope that biochemical capability to fix atmospheric nitrogen may eventually be transgenically implanted into cereal crops.

OTHER SOURCES OF NITROGEN

Other nitrogen sources include animal manure; bio-fertilisers such as leaf mulch; compost from crop residues, in some instances assisted by earthworms as agents of conversion; green manure from leguminous plants and trees; such aquatic organisms as azola and blue-green algae. Manure, compost, mulch and other bio-fertilisers provide organic and fibrous materials beneficial to soil structure. Nutrient concentrations vary and rates of release from bio-fertilisers are not easily controlled, being influenced by soil temperature and moisture content. Release of nitrogen is generally slower from animal manure and composted residues high in lignin than from chemical fertilisers.

It is difficult to collect, store, distribute and utilise animal manure without significant nitrogen loss resulting from liquid run-off, leaching by rain, and volatilisation of ammonia. Manure stored in the open air in humid tropics with high rainfall can lose 70 per cent of nitrogen in 24 hours. Manure from cattle driven by pastoralists in semi-arid regions quickly dries and is blown away. Where wood is scarce and kerosene is expensive, animal manure may provide the cheapest source of household fuel. In contrast, chemical fertilisers are stable until used, designed and compounded to release nutrients as specifically needed in desirable composition and concentration.

Nevertheless, excessive application of chemical fertilisers may cause soil degradation, toxicity to earthworms and other useful soil organisms, and by leaching into ground and surface waters may cause insult to human health and harm to fish stocks. In northern European countries, where high cattle stocking densities are common, serious nitrification of ground water results from manure accumulations that freeze during the winter and release high concentrations of water-soluble ammonia during the Spring thaw. Farmyard manure is a source of various coliform microorganisms some of which are pathogenic. An outbreak of listeriosis in Canada was traced to raw salad vegetables fertilised with farmyard manure.

3

Modern Agriculture in Farming Management

PRIMITIVE AGRICULTURE

Historical writers agree that in early days, the inhabitants of the earth gained a livelihood by hunting and fishing. When the population of a region became so dense as to deplete the supplies of food to be gained in this way, the more enterprising ones tamed and domesticated some of the wild animals for their own use. These animals were given protection and provided with food so that they multiplied rapidly. In this way a supply of meat was insured for those who cared for and protected them. From hunting and fishing the people turned to such pursuits as grazing and herding, and gave much of their time to caring for animals. This was known as the pastoral stage of subsistence. It was more reliable than hunting and fishing for several reasons. By protecting the domesticated animals from the beasts of prey and driving away the less useful animals, larger numbers of useful animals were enabled to live in a given region. Food supplies of meat were therefore more plentiful and certain.

This custom of providing food supplies for animals led to the discovery that certain plants were better adapted for food than others. To raise these desirable plants in abundance it became necessary to destroy other plants that were not so useful, but which contended with the useful plants for possession of the soil. This may be said to be the beginning of agriculture. At least it was the first indication of the organization of nature's forces to meet the needs of man. The demand for land for grazing and tilling purposes led to mutual agreements for territory which should be used for such purposes by certain persons or tribes, and is the first suggestion of land allotment or ownership.

British Agriculture

The agriculture of England has passed through many different stages. Because of scarcity of agricultural labourers between 1350 and 1375, many of the owners of large estates were forced to change their system of farming. Previous to that time most of the tillable land had been devoted to grain raising, bread being the chief article of food. Then the land was changed into pasture, and sheep and cattle raising took a prominent place in British agriculture.

The greatest development came about the middle of the seventeenth century with the introduction o fclover and turnips as field crops. The chief advantage in these crops was the increased amount of feed, which enabled the farmer to keep more cattle on his land. Keeping of cattle gave manure that was applied to the land and resulted in better handling of the farms. The turnips took the place of bare fallowing which had previously been practiced, thus enabling the farmer to use all of his land each year.

Improved methods of tillage followed rapidly. About the first of the eighteenth century Jethro Tull started the practice of drilling the grain in rows, later advocating tillage between the rows, and inventing implements for the work. Lord Townshend, about the same time began the rotation of crops, greatly improving the producing power of his land thereby. Another notable man in British agriculture is Coke of Holkham, who like his predecessors grew turnips and clover, but who also studied economical management of his farm. He reduced the number of horses used in plow teams and increased greatly the efficiency of his live stock. Following the work of these three men, farming became a gentleman's profession in England and progress since has been rapid. The development and improvement of live stock has received so much attention at the hands of British breeders that the whole world practically has been drawn to their markets for foundation stock.

American Agriculture

The agriculture of America has been developed largely from the methods followed on the British Isles. The American colonists brought with them the customs of the British farmers and have since gone frequently to the mother country for improved seeds, for live stock, and even for men to do certain parts of the farm labour, as for instance, caring for the more valuable imported stock.

The development of agriculture during colonial days was slow and full of hardship. The British and European methods of agriculture required re-adapting to the new conditions. The people were forced to live a new kind of life, to clear land of trees and stones, and to learn by experience what crops were adapted and what methods of tillage would succeed. From the Indians they learned how to grow corn and tobacco, which have ever since been two of our important crops. From them also they learned many of the secrets of soil and climate. The agricultural development proceeded along two lines.

In the South large plantations with negro slaves as labourers were established. Such establishments were practically self-sustaining because they supplied nearly all articles of food and clothing. In the North the family farm was developed largely and the labour of the farm was performed mainly by the farmer and his family. The conditions existing in the North encouraged the cooperative spirit and developed sociability among the people. Land in either section was abundant and the greatest problem in the management of farms waws that of economizing labour.

The colonial development was along the eastern sea coast. Following the Revolutionary war, settlement moved westward beyond the Alleghany Mountains and into the Great Central Basin. The invention of the cotton gin about 1793, and the development of the cotton industry, changed rapidly the systems of farming in the South. Slavery became greatly extended to provide cheap labour for the cotton fields, and probably had much to do with the rapid settlement of the southern states.

Cotton and tobacco growing were the leading industries in these states and agricultural improvement was slow because labour was cheap as compared with that in the northern states and the use of improved machinery was not so imperative. Migration westward was encouraged by the opening of the Northwest Territory in 1785.

A liberal land policy on the part of the National Government and the discovery of gold in California served to stimulate the movement westward and aided the rapid settlement and development of the great prairie regions of the Mississippi Valley. The invention and perfection of farm machinery, begun about 1825 to 1830 and continued to the present time, has also been a great factor in the development of agriculture in America. The reaper, invented, the thresher, perfected about 1850, and steel plows for turning the soil, manufactured since 1840, have had a great influence on the agriculture of the northern states which were especially interested in grain and corn raising.

Live stock raising was followed to some extent in the northern states also, but received great impetus about 1839 by the importation of the English thoroughbred stallion "Denmark" into Kentucky. Percheron draft horses from France were also introduced soon after. Corn raising on the western prairies stimulated greatly the growing of hogs. Cattle raising on the vast prairies also was greatly extended during this period.

Dairying began to develop about 1850. The improvement of machinery and the increase in stock and grain raising went on steadily in the northern states, not being interrupted seriously even by the Civil War. In the South where cotton and tobacco grown on large plantations by slave labour were the main crops, agriculture was seriously demoralized by the war which set free the slaves, thus destroying the source of cheap labour.

Several years were required to build up a new system of agricultural production and to adjust labour problems, but the agriculture of the South is now on a much better basis. During the past twenty-five years agricultural development has been particularly rapid. The establishment of experiment stations in 1888 in every state in the Union, through the passage by Congress of the Hatch Act in 1887, and the development of Agricultural schools and colleges has led to a systematic application of science to agriculture. Crop growing has become much more certain and live stock growing has been largely developed.

The occupation and settlement of the prairie lands, the perfection of machinery, and extended opportunities for agricultural education, have been large factors in improving conditions for farming. In the days of abundant land, large holdings were secured by many farmers. The tendency in later years is to reduce somwhat the size of farms and to take up more intensive forms of agriculture. Greater efficiency is demanded in the management of farms in order that greater profits may be realised by the tiller of the soil.

FARMING MUST BE ORGANIZED

During the period of cheap land and comparatively cheap labour, many crops were raised which do not under existing conditions, yield a profit. Methods of tilling the land and handling the crops that were followed during that period would not now give large enough yields to support the farmers and their families. Crops and methods of tillage must be readjusted to meet the changed economic conditions and satisfy the demands of the rapidly growing population for food stuffs.

The practice of grain raising so prevalent in many portions of the United States must give way to live stock raising, fruit and vegetable growing and other more intensive systems of agriculture. A great variety of crops and a larger supply of all food stuffs and clothing material is required by the present population. The value of the agricultural products in the United States in 1910 was $8,694,000,000. The total value of all live stock on farms in the United States in 1910 was $4,925,000,000. The demand is constantly increasing. The future supply of these commodities is a matter demanding the most serious consideration.

It is important that the old soil robbing types of farming be superseded by more conservative methods; that the land be so tilled that it will bring the largest crops at the least expense of soil fertility; and that these crops be used or fed as near as possible to the land upon which they were grown, the manures and crop residues being returned to the soil.

It is important also that the farmer receive a fair compensation for his labour. Crops must be selected that pay the best profits. These will differ in different years and localities. The farmer, therefore, must learn to select wisely and manage well so as to make a reasonable profit on his labour and investments.

The business of farming is constantly becoming more complex. The keen demand for land, the scarcity of labour, the desire for financial gain and the call of the city residents for supplies are acting as a spur for still greater effort. To meet these demands a new science is being developed which aims to correlate the various factors entering into the farmers' business so that it will be certain and remunerative. An effort is being made to organize the business of farming so that it may provide for the needs of the present generation without exhausting the source of supplies for generations yet to come. This new science is called *Farm Management*.

MODERN TECHNOLOGICAL NEEDS AND FUTURE AGRICULTURE

It is apparent that the tasks of meeting the consumption needs of the projected population are going to be more difficult given the higher productivity base than in 1960s. There is also a growing realization that previous strategies of generating and promoting technologies have contributed to serious and widespread problems of environmental and natural resource degradation. This implies that in future the technologies that are developed and promoted must result not only in increased productivity level but also ensure that the quality of natural resource base is preserved and enhanced. In short, they lead to sustainable improvements in agricultural production.

Productivity gains during the 'Green Revolution' era were largely confined to relatively well endowed areas. Given the wide range of agroecological setting and producers, Indian agriculture is faced with a great diversity of needs, opportunities and prospects. Future growth needs to be more rapid, more widely distributed and better targeted. Responding to these challenges will call for more efficient and sustainable use of increasingly scarce land water and germplasm resources.

Technical solutions required to solve problems will be increasingly location-specific and matched to the huge agroecological/climatic diversity. Detailed indigenous knowledge and greater skills in blending modern and traditional technologies to enhance productive efficiency will be more than ever before, key to the farming success and sectoral growth. Most technological solutions will have to be generated and adapted locally to make them compatible with socio-economic conditions of farming community.

New technologies are needed to push the yield frontiers further, utilize inputs more efficiently and diversify to more sustainable and higher value cropping patterns. These are all knowledge intensive technologies that require both a strong research and extension system and skilled farmers but also a reinvigorated interface where the emphasis is on mutual exchange of information bringing advantages to all. At the same time potential of less favoured areas must be better exploited to meet the targets of growth and poverty alleviation.These challenges have profound implications for products of agricultural research.

The way they are transferred to the farmers and indeed the way research is organized and conducted. One thing is, however, clear – the new generation of technologies will have to be much more site specific, based on high quality science and a heightened opportunity for end user participation in the identification of targets. These must be not only aimed at increasing farmers' technical knowledge and understanding of science based agriculture but also taking advantage of opportunities for full integration with indigenous knowledge. It will also need to take on the challenges of incorporating the socio-economic context and role of markets.

With the passage of time and accelerated by macro-economic reforms undertaken in recent years, the Institutional arrangements as well as the mode of functions of bodies responsible for providing technical underpinning to agricultural growth are proving increasingly inadequate. Changes are needed urgently to respond to new demands for agricultural technologies from several directions. Increasing pressure to maintain and enhance the integrity of degrading natural resources, changes in demands and opportunities arising from economic liberalization, unprecedented opportunities arising from advances in biotechnology, information revolution and most importantly the need and urgency to reach the poor and disadvantaged who have been by passed by the green revolution technologies.

Another important implication of increasing globalization relates to the need for greater attention to the quality of produce and products both for the domestic and the foreign markets. This would imply that production must be tuned to actual rapidly changing product demand. Such adaptation to global markets would require state of the art research, which can be achieved only by setting global standards of research, focus on well defined priorities and mechanisms which permit close interaction of farmers with researchers, the private sector and markets.

CHALLENGE OF INDIAN AGRICULTURE

There are several concerns about Indian agriculture that we need to address. Even in the green revolution crops, rice and wheat, the Indian average yield ranks around 50th in the world. A weak agro-based industry, the problems of weak marketing, storage, transportation, credit support, etc, haunt us. Only 0.3 per cent of agricultural GDP is spent on agricultural R &D. The population is growing at an alarming rate of 1.8 per cent per annum. It will be around 1.3 billion by 2020 and would require 50 per cent additional foodgrains. A total of 200 million people are still below the poverty line and have insufficient access to food. Declining resource base, degeneration of natural resources, including soil fertility, water environment, etc, are some other concerns. Food production systems centre around a few states only. Out of 17 big states, only 7 are self-sufficient and surplus in foodgrains. There was a significant decline in per capita intake of calories and protein between 1975 and 1990.

Inspite of all these problems, there is plenty of good news. India harbours a vast diversity of basic natural resources. Tremendous bio-diversity of agriculturally important micro-organisms exists, which are not yet been harnessed. We have a variety of crops, including quality fruits, vegetables, foodgrains, and a number of plantation and commercial crops. Indian fisheries resource has spread over 8,100 km of coastline on the east and west, and provides access to marine resources and transportation.

Very large proportion of area of various food crops is under low productivity category. It can be increased with inputs of high class science. There is a scope to cultivate sizeable portion of waste lands through soil amendment and introduction of agroforestry. Several million hectares of saturated soil can be potentially exploited from rainfed lowlands of eastern India.

The India Exclusive Economic Zone (EEZ) of the marine sector, of about 2 million sq.km., has high potential harvestable yields. Only about 20 per cent of the cultivated area is covered under tractor cultivation, great scope of making quantism quantum jump in production and productivity exists, changing consumption patterns offer scope for diversified agriculture. Trade liberalization provides opportunities to reach for outside markets, which were not accessible otherwise. Great gains are possible through agro-based industrialization and overall rural development.

The cultivators and farmers of India have a rich heritage of traditional agricultural wisdom on crop and livestock husbandry and fisheries; based on the principles of the conservation of natural resources and environment. India has an extensive network of extension channels for dissemination of technologies generated by research institutes. India has a strong manufacturing capability and a large network of about 20,000 manufacturers and nearly one million village artisans for indigeneous production of all types of agricultural machinery. Therefore, inspite of several disadvantages, we have a great chance to leap forward, if we enforce the right policy initiatives and implement them.

BIOTECHNOLOGY AND INDIA AGRICULTURE

We need to recongnise that, just as the food reequirements of today's population of nearly 6 billion people could not have been met by the technologies of the 1940's, we cannot assume that current practices will feed the population of 8 billion expected by 2020. New approaches are needed in addition to the continued improvement of existing methods of crop and animal husbandry and food processing.

The strategic integration of biotechnology tools into India agricultural systems can revolutionise Indian farming and usher in a new era. Genetic engineering is clearly the most revolutionary tool to impact agricultural research since the discovery of genetics by Mendel. Prior to genetic engineering, the exchange of DNA material was possible only between individual organisms of the same species. With the advent of genetic engineering in 1972, scientists have been able to identify specific genes associated with desirable traits in one organism. For example, a gene from bacteria, virus or animals may be transferred into plants to produce genetically modified plants having changed characteristics. This method therefore allows mixing of the genetic material from species that cannot otherwise breed naturally.

Genetic engineering can be vital for an agrarian country like India. It can help in minimising the crop damage through disease and pest resistant varieties, reducing the use of chemicals, enhancing stress tolerance in crop plants, thus permitting productive farming on unproductive lands, etc. One could even extend the growing season of crops and minimise losses due to environmental factors on the one hand and increase the shelf life of fruits and vegetables, on the other, thus minimising losses due to food spoilage. We can expand the market vista and improve food quality. Biotechnology can also produce plants that possess healthy fats and oils, possess increased nutritive value, and create a whole range of higher value feeds. Biotechnology has even the power of producing biodegradable plastics, edible vaccines, etc.

It is only through the blending of the 'gene revolution' with our experience in the 'geen revolution' that we can reach our goal of 'evergreen revolution' and also 'nutritional revolution' The advantage of the gene revolution is that it is relatively scale neutral, benefiting big and small farmers alike, It is also environment friendly. Thus, it can be of great help to the smallest farmer with limited resources in increasing farm productivity through the availability of improved but powerful seed. It can also reduce a farmer's dependency on chemical inputs such as pesticides and fertilisers.

Modern biotechnology offers unlimited opportunity for enhancing genetic potential of crops and other commodities, management of biotic and abiotic stresses, bio-remediation and organic recycling. India is one of the main centres of agricultural biodiversity and its gene richness can greatly complement the developments in modern biotechnology. Likewise, new developments in GIS, remote sensing, and crop modelling, provide new opportunities for integrated management of natural resources. The revolution in informatics provides opportunities for sharing latest information for research and planning in a highly organised and efficient manner. With the globalisation of economy, the opportunities for value addition and post harvest management are also immense. With clever blending of technologies, the India farmer can usher in new era of confidence and performance. Sir Francis Bacon had once said: *'It would be an unsound fancy to expect that things which have never yet been done, can be done except by methods which have never been tried'* In the new context, we have to review the new role of biotechnology. Also, it is in this context that we have to view the rapid advances that modern biotechnology is making, including the area of transgenics. Let us focus on some of the developments in this area, since considerable interest and controversy has been created around the world today in this exciting area.

MOVEMENT IN TRANSGENICS

The advent of genetic modification over the last two decades has enabled plant breeders to develop new varieties of crops at a faster rate than was possible using traditional methods, with huge potential for further beneficial development.

Tobacco was the first plant to be genetically transformed\, in 1983, with cereals beginning in 1990, but it is only recently that products such as GM soya have reached the market place in Europe, and, for a variety of reasons, given rise to concern & controversy.

The concern must be addressed seriously if society is to exploit the new technologies appropriately. We will address this issue rather extensively a little later. Whereas in India, we have been discussing and debating the issues, the rest of the world is galloping ahead.

Table. Ag Biotech Product Sales

$ Millions		**Sales**		**Annual Growth Rate**	
	1994	1997	2002	1994-97	1997-02
Transgenic seeds and plants	20	405	2,030	173%	38%
Animal growth horomes	80	225	405	31	12
Biopesticides	35	65	110	23	11
Other	107	180	340	19	14
Total U.S. Sales	$242	$875	$2,885	53%	27%

Development, field testing, and commercialisation of transgenic crops, is now recongnized by the international scientific community to be an essential strategy for food security. Fortyfive countries, including India, have conducted transgenic crop field trials up to 1997. The activity in the developing countries has been shown in Table below.

Table. Releases of Transgenic Plants in Developing Countries (by Species and Introduced Trait)

Species			**Introduced Trait**			
	Total Fied releases	**Herbicide Resistance**	**Insect Resistance**	**Virus Resistance**	**Product quality**	**Others**
Maize	46	29	16	1	3	5
Soya bean	27	25	—	1	1	—
Cotton	24	16	15	—	—	—
Tomato	19	—	2	1	16	—
Potato	13	—	1	6	—	6
Subtotal	129	70	34	9	20	11
Other species	30					
Total	159					

The emphasis has been mainly on traits for herbicide, insect, virus & fungal resistance, and on product quality with a limited number of genes. Out of a total 159 releases. Argentina leads the way with 43, followed by chile, Mexico, Puerto Rico and the Republic of South Africa, with 17-20 each. India is far behind with just a few releases. There are fears relating to transgenic plants in society. These need to be addressed, not by brushing them aside, but by careful scientific analysis and educational programmes. All biosafety protocols must be followed rigidly. Let us deal with this issue more exhaustively now.

PUBLIC CONCERNS

The potential of biotechnology as a method to enhance agriculture productivity in the future has been accepted globally. However, because of its revolutionary nature, risk and uncertainty may be created by the process of genetic engineering and by the resulting genetically modified products. This may result from the introduction of a new unrelated DNA sequence into a recipient organism. The introduced DNA may have some unexpected effects on the cellular processes of the recipient organism. In addition to this, introduction of antibiotic resistance genes, as selection markers, pose serious implications in public health especially in genetically modified plants that are directly used for food purposes.

Risks are also associated with the genetically modified plants that are released into the environment. The nature of interactions with other organisms of the natural ecosystems cannot be anticipated without proper scientific testing. For example, modified plants with enhanced resistance to pests or disease threaten to transfer resistance to the wild relatives, which will have implications for biodiversity and ecosystem integrity. These, and many more, doubts plague the minds of common people. Therefore, it is very clear that if the biotechnology potential has to be completely realized, it has to be done in an extremely responsible manner. At each step, proper testing and scientific data have to provided. Not only this, proper regulatory and policy mechanisms have to be put into place.

It is therefore, absolutely essential, that conscious efforts be made especially by the developing countries to initiate well-defined programmes for the development and regulations of genetically modified plants. To make this possible and to benefically, ethically and sustainably reap the benefits of biotechnology, there is a pressing need for scientists, researchers, policy makers, NGOS, progressive farmers, industrialists, and representatives of the government, to come together on a common forum in order to discuss the common concerns and find solutions to them.

Will modified plants transfer their introduced genes into wild relatives growing nearby? Will modified plants that produce new compounds disrupt the 'balance of nature' in some way?

Could the planting of a restricted number of cultivators leave crop plants more susceptible to disease? Could the planting of a restricted number of cultivators lead to a reduction in biodiversity (of crop plants, weed species, insects and microflora in the fields in question)?

Will genetically modified plants be able to avoid the factors that regulate natural populations and thereby change the usual ' balance' between populations?

Similarly, there are concerns about the social and economic effects. For example, some of these concerns are:

- How will the structure of farming (particularly in developing countries) be affected by biotechnology?
- How will patent laws affect traditional breeders 'rights (*e.g.* the right to save seed from one year to the next)?
- Will plant breeding be left increasingly in hands of a few companies, and if so, what effects might this have?
- Will some countries be 'plundered' for their genetic resources?
- We also have a set of ethical and moral issues that need addressing. In particular,
- The consumer has the right. What will consumers be told about the new food products?
- Is it acceptable to 'interfere' with nature through genetic engineering?
- Do we have the right not to use all means available to improve crop plants, especially when so many people are under or malnourished?

Then there are several regulatory issues. Among these, the three key issues are:

- Do current regulations give sufficient protection to farmers, consumers, those who have invested in, and those engaged in research?
- Is there sufficient international legislation to ensure environmental protection?
- Do current regulations compromise the competitiveness of biotechnology companies by being excessively restrictive?

The Royal society of London had appointed a group of experts to examine various aspects including the scientific evidence concerning the risk of transfer of genes from Genetically Modified (GM) crop plants to wild species and non-GM crops, the uptake of genes from GM food by the digestive system, and the current state of the regulatory system.

The experts concluded that the chances of gene transfer happening are slight, provided the regulatory processes are followed, but that this must be kept under consideration.

The experts also concluded that the uptake of genes via the food chain is not a new issue because genes (*i.e.* DNA) are normal constituents of the human diet. Many products from GM Plants, such as sugar prepared from GM tomato paste, are so similar that they are regarded as 'substantially equivalent' others, for example flour from GM soya, may contain a new gene or its product, although many of the purification processes involved in food production will destroy and DNA present in the raw material.

Some GM foods have been produced using an antibiotic resistance 'market' gene, which is a laboratory device designed to identify genetically transformed plants.

The Royal Society has shown concern about the use of such genes in food products and it suggested that any further increase in the use of such markers in the human or animal food chain would be undersirable. Gm crop plants have been produced to improve insect tolerance and virus resistance, and to include herbicide tolerance, so that transferred to non-target species of plants, and that the development of resistance by target pests is minimised, the regulatory authorities must be assured that any negative effects would be no greater than those resulting from conventional procedures; and that any long term effects on the environment and ecology would be closely monitored, with statutory restrictions in place to control marketing; and also that best practice advice was adopted by growers.

The reliance on a case by case approach to the legislation, may result in lack of analysis of the overall impact of the technology on agriculture and the environment, and of the long term effects of gm. Although mechanisms are already in place to regulate many individual aspects of GM technology, there is no means for looking at GM technology as a whole. As independent, overarching, regulatory body is needed to span departmental responsibilities, monitor the enforcement of existing or future regulations, and strengthen the guidelines to growers of GM crops, such as those specifying the isolation distances between Gm and non-GM crops. The body should also review and monitor the membership of advisory committees and regulatory bodies.

BUSINESS SIDE OF FARMING

FARMING AS A BUSINESS

Farming has not usually been considered a business. The diversity of the duties of the farmer, the area over which the operations of the farm extend and the complexity of the records required, combine in making difficult the organization of the details of farming into business form. The successful financial operation of a farm presents quite as complex problems and calls for at least as much business ability and judgment as is required in operating a store with the same investment.

Farming, therefore, should be considered as a business, and the man who can produce his crops and products at the lowest cost and sell them at the highest price, investing the proceeds to the best advantage, should be considered the best farm manager. The man who knows the details of the cost of production and operation, and whose records show the profitable and unprofitable lines of production, thus enabling him to eliminate those that do not yield a profit, may be counted as the best business man.

A farmer should know the elements of soil fertility. He must understand the principles of the movement of soil water, and the action of soil bacteria. He should understand the nature of plant growth and be familiear with varieties and species of plants and with the effect of one crop on the crop following. He must also be familiar with animals and their habits and know how to feed and care for them. In addition, he must know how to buy and sell to advantage, make contracts, and plan his buildings and his farm so as to necessitate a light expenditure for labour, also that he may distribute his labour to advantage over the various farm enterprises. And he should know how to keep accounts.

The farmer in organizing his business could well follow the example of the merchant. The merchant first takes an inventory of his stock. He studies the demand for his goods, both present and prospective. He notes the supply, the cost, and the demand for each article. He calculates the labour required to operate his business and such other items of expense are considered as may be legitimately charged against the business.

He regulates his purchases and his prices according to the cost of securing his goods and putting them on the market. In conducting a large store business, it is customary to organize it into departments, putting some competent person in charge of each department and having the labour and accounting charges so systematized and recorded as to show the profit or loss from each department and from the business as a whole.

The farmer should likewise take an inventory of his capital, stock, and equipment. He should consider the type of farming to which the soil and climate are adapted. He should consider the fertility of the soil and the demand that will be made upon it by the crops grown. He should consider, in connection with the soil fertility, the sources from which it may be renewed and at what cost. He must study the markets, the transportation, and the demand for such crops as he grows; also the cost of producing each of the crops and the probable net profit that will be returned. His labour likewise should be charged against the various crops or enterprises and distributed to the best advantage.

In studying the problems of farm organizations, interest on investment, taxes, insurance, and other expense must be included as they affect the financial result. As in a large store business, it is frequently necessary to organize the large farm into departments, keeping accounts with the dairy, with the swine, the grain crops, the garden, and other similar enterprises.

Where the business is large enough, it is well to put an expert in charge of each large branch or group of enterprises, thus enabling one to use cheaper labour for performing the work or making the labour more effective by closer supervision. Where the farming is conducted as an organized business, and accounts are kept with the various lines of work, it is possible at the end of the year, to make a business statement which will show which lines have been profitable. The manager then can change his methods or drop out those lines that prove to be unprofitable and the business as a whole may be put on a better basis.

INVESTMENT

The investment of money in land, buildings, and equipment demands careful consideration. It is possible to pay so much for a farm that it will be impossible to produce sufficient revenue to meet the expense of operation and to pay a normal rate of interest on the money invested. This is particularly true where low-priced products are produced. A farm may be highly productive but so located that it will be impossible to market the produce on a profit bearing basis. One should study closely the market facilities of the neighbourhood and raise supplies which can be successfully marketed locally, or which can be transported to a market that pays well for such produce. Unless the produce of the farm is well related to the market, the farm is likely to be operated at a loss.

PROPORTION IN REAL ESTATE

It is a mistake for one to invest all of his capital in the real estate itself. Sufficient capital should be reserved for operating the farm. The hunger for land has induced many farmers to buy more land that they can equip and operate well. Such farmers are said to be land poor. They would secure greater profit from a medium sized farm, well tilled and managed, than from a large one which is insufficiently equipped and poorly operated. Rarely should more than 50 or 60 per cent of the capital be tied up in the land. The size of the farm and the amount of equipment are closely related to the possible profits.

A farm of forty or eighty acres devoted to diversified crops and live stock, cannot be so economically equipped per acre as a larger farm. The investment per acre in machinery will be higher as the cost will be spread over fewer acres than the machinery has capacity to handle. Investment in other equipment will also be correspondingly high. Often the labour on such a farm is not fully employed and loss results from inactivity of labour and equipment. A medium to large sized farm, when well organized, fully equipped, and with sufficient capital reserved to operate it well, will pay a much better labour income than a small farm.

PROPORTION OF INVESTMENT IN MACHINERY

Investments in machinery are worthy of quite as much consideration as investments in land. Machinery is looked upon as one of the means of reducing the cost of production. The wise use of machinery saves time and labour and enables the farmer to handle large acreages. In this light, the use of ample machinery is wise. The fact remains, however, that investments in machinery are often poorly made and that many farmers are embarrassed by debts for machinery. Frequently, farmers purchase machinery because it is fashionable or because a neighbour has it, rather than because carefully made calculations show that a certain machine can be used profitably. A safe rule is to buy no machine until carefully made calculations show that the cost of production of a certain crop or product will be reduced sufficiently by the purchase to cover the cost of the machine. That machinery investments can be studied from a business standpoint is quite plain. The following example, showing the relative cost of cutting corn with a machine versus cutting by hand, will illustrate:

The original cost of a corn binder is $125. The annual depreciation, as shown by statistical records covering ten years' work with farmers in Minnesota, is $12.50. The interest on the investment at the average value of the binder throughout its life will be $4.12. Repairs, shelter, and insurance will cost $2 annually. The total annual cost for the use of the binder, therefore, will be $18.62. If only twenty acres of corn are grown each year, the annual cost an acre for the corn binder will be $.93.

The cost of cutting corn would be as follows: It will be noted that a twenty-acre corn field can be more economically harvested by hand where labour is available.

On a ten-acre field, the difference in expense would be still greater in favour of the hand harvesting, as the cost of machinery per acre would be doubled. The scarcity of labour, however, and the necessity of harvesting corn quickly to save it from frost, would often warrant the expenditure for the machine, even though it does slightly raise the cost of harvesting per acre. The greater the acreage, the more useful the machine becomes in harvesting, and the less the expense per acre for machinery use. Calculations similar to this should be made before purchasing a machine for any purpose. If it can be shown that the cost of performing the labour may be reduced by the machine, and that the labour can be performed in a more satisfactory and efficient manner with the use of it, then the purchase may be warranted. In many cases, however, calculation will show that the purchase is not warranted and that it would be better to hire labour and rent the machine, or to buy in partnership with some one else.

COST OF MOTIVE POWER

Another factor that should receive consideration is the cost of motive power in use on the farm. Horses usually furnish the farm motive power, though tractors are used to advantage in some cases. Auto trucks and automobiles can often be used to advantage in marketing dairy, fruit and garden products and the cost of using them, when it can be ascertained, should be compared with the cost of horse power. It costs from $40 to $100 per year to keep a work horse, depending on the locality and on the price of feed stuffs; also on the work that the horse does. The average cost a year of keeping a horse in Minnesota for the years 1904 to 1907, varied from $75.07 at Halstad, to $90.40 at Northfield.

Often a large number of horses are kept because the farm has been devoted to grain raising and the horses are needed at seeding and harvest times. They run in the pasture during the summer and are idle during the winter months. They must be fed and cared for during this time. The money invested in them would be drawing interest if invested somewhere else. Farmers should reduce the horses kept to the number actually required to do the work, unless the surplus are colts growing in value as one of the market products of the farm. The work of a farm can often be lessened by adopting a good crop rotation and using such crops as do not demand large amounts of horse labour at the same time. In this way a farm of 160 to 240 acres can often be worked with four to six horses, whereas eight to ten are frequently kept. The support of two or three extra horses per year would amount to $200 to $250 and is an item well worth saving.

INVESTMENT IN BUILDINGS

Investment in buildings, fences, and other items of equipment should be considered in the same business-like way. A barn costing $4,000 and providing shelter for forty head of cattle would carry with it an annual cost of $440. This annual cost is made up from the interest on the investment, insurance, depreciation, paint, and repairs. It will be about 11 per cent on the total investment. If the same forty cattle could be housed in a barn costing $2000 the total annual cost would be only $220, charging the same interest and expense rates as in the first instance, and assuming that the rate of depreciation would be the same.

While the $4000 barn would undoubtedly be a better barn, it would not add to the production of the cows housed, unless it was much more comfortable. It would not add to the net profit from the investment unless the labour of doing the chores and caring for the cows would be considerably reduced by greater convenience. The cost of horse barns, swine barns, and other buildings can be similarly calculated. One should not erect a building unless it is going to add to the efficiency of the live stock, shelter hay or machinery, or lessen the labour of doing the chores. It is wiser to invest money in drainage or in better tillage of the soil, than to invest it in buildings that shelter unproductive stock, or that add nothing to the earning power of the farm.

COST OF LABOUR

The employment, organization and direction of labour demands considerable study. The value of a good farm manager lies quite as much in his abilitiy so to select and direct labour as to yield a profit, as it does in his ability to drive a good bargain or sell his crops well. The only reason for employing labour is to increase the product and consequent profit. If a farmer can, by employing a man eight months in the year at $40 per month, increase the product of his farm by $500, he will be warranted in employing the labour. If, however, the $320 invested in labour should yield an increase of only $200 in the products of the farm, employment would be at a loss.

FACTORS OF PRODUCTION

Three primary factors are necessary in agricultural production. These are capital, land, and labour. The adjustment of these three factors is an important part of the business of the farm owner or manager, and determines largely the profits that may be made from the individual farm.

Capital, as commonly understood, includes the money value represented in the investment of the farm property, no matter what the form may be. Implements, live stock, teams, buildings, and othe rarticles of equipment, ar each a part of the capital of the farm. Cash for operating is also included.

*Good buildings and fences and a well kept farm often help in attracting customers for stock, seed grain or other products. As an advertisement they may increase the earning power of the farm indirectly. The satisfaction of owning good buildings and their influence in keeping the young folks on the farm or in enabling one to keep hired help should also be considered. The reputation of a farm in this respect may become a business asset.

Land represents the larger part of capital on most farms, and demands special consideration because the amount of land available for agricultural purposes is limited. Farmers have for this reason, regarded it wise to secure large quantities in localities where it was cheap, anticipating a rise in value. Location and demand for land in particular sections has led to much speculation, and land values fluctuate frequently.

The proportionate investment in each of the three forms,—circulating capital, land, and labour—bears a vital relation to the profits possible from the farm, and must be given the most careful consideration by the person who is buying and equipping a farm.

CAPITAL CLASSIFIED

There are two forms of capital in common use. They are known as fixed or invested capital, and circulating or working capital. The fixed capital properly includes all forms of permanent equipment, such as investment in land, buildings, implements, teams, and other articles that are used continuously. In land it includes the natural value and the value of the improvements that have been made upon it. Picking stones from a rough section of land adds to its value and increases the capital invested. Clearing trees from the land has the same effect. Wells, drainage, roads, fences, and other forms of improvement which are permanent and which become part of the land, also add to the natural value and become a part of the fixed capital. Buildings also are looked upon as part of the fixed capital. Strictly speaking, only those buildings which add to the producing power of the farm should be included in the capital invested in the farm. The dwelling house, while commonly added to the investment in the farm, is really intended for the personal use of the farmer and his family. Except in so far as it shelters the help employed on the farm, it can add but little to the returns from it. So far as making a statement of the business of the farm is concerned, it would be better were the farm-house inventoried separately from the other buildings and regarded as a personal expense to the farmer, just as the house of the banker in the city is separated from the business of the bank. All other buildings including silos, corn cribs, granaries, and buildings for sheltering the stock and necessary in conducting the farm business, should be included in the inventoried capital of the farm.

Equipment in the way of teams for work purposes; implements; live stock, such as cows, brood sows, sheep and poultry, that are kept for live stock products, are all a part of the permanent equipment, since they are permanently employed and if sold are replaced by other animals for the same purpose. The circulating or working capital, includes such items of equipment as are frequently changing. Seed grain, household and farm supplies that are immediately used or marketed, live stock, such as fattening steers, and money for hired labour, are examples of circulating capital. The classificiation intends tha the term "circulating capital" shall include only those items that are used once and disappear. If sold for cash, the cash may be invested in other forms of working capital which in turn disappear. Needless to say, the amount of working or circulating capital varies greatly in accordance with the type of business done, with the market, and with the tastes of the farmer. No rule can be given for the exact adjustment of capital for these reasons.

PRODUCTION LIMITED BY DEFICIENT FACTOR

It is a common experience that the production on a farm is limited by the minimum amount of the one deficient factor. Difficulty is experienced in securing profitable production on a limited land area. In such a case, labour will not be fully employed or the equipment cannot be used effectively. On the other hand, a large land area and large equipment cannot be used to advantage without a plentiful supply of labour. Again, neither land nor labour can be used to the best advantage if the equipment is inadequate. It therefore stands to reason that these three factors must be carefully considered and proportioned in accordance with the needs of the business.

In purchasing and organizing a farm with limited capital, it is believed best to make the investment in about the following proportions:

- 45 per cent for land investment
- 20 per cent in buildings, provided they are to shelter productive live stock or market products;
- 22 per cent in work animals and live stock;
- 8 per cent in implements and tools;
- 5 per cent reserved as working capital.

In buying a farm and equipping it with new machinery, the land investment will run lower than above indicated and the implements and tools investment will run higher. The machinery, however, will depreciate in value while the land is more likely to increase in value.

As the farm becomes older and the land is improved, the proportion of investments will gradually change. On old farms near city markets, the machinery investment may become comparatively insignificant.

EXERCISES FOR PUPILS

Farm Inventory

Have the pupils take inventories of their fathers' farms, providing a form similar to the following. The inventory may be taken on the regular weekly holiday. Prices should be checked over by the proprietor of the farm and brought to class for discussion and completion. The teacher should, if possible, go with the class to some farm and take an inventory, with the help of the proprietor, before starting the class members on their individual work. Allowing for any withdrawals of or additions to capital, the difference between the totals of two successive inventories will be the gain.

Proportion of Investment

Have the pupils find the total investment and the per cent of the capital invested in each of the forms of equipment used in classifying the inventory.

Cost of Shelter

Have the pupils learn the cost of the barns on their fathers' farms. Ask them to calculate the interest at the prevailing rate. Determine the amount of depreciation on each barn at 5 per cent. Learn the cost of insurance and repairs for the year. Include the interest, depreciation, insurance, and repairs in one sum called the "Annual Cost." Divide this annual cost by the number of animals sheltered and learn the cost of sheltering one head on each of the farms. Where different animals are kept, they may be reduced to a comparative basis by estimating the weight and considering 1000 pounds as a unit.

ECONOMY OF LANDLORD FARMING IN THE AREA DESCRIBED

We have cited the evidence of agronomists and farmers to the effect that dairy farming on the landlord estates leads to the rationalisation of agriculture. Let us add here that the analysis of the Zemstvo statistics on this question made by Mr. Raspopin fully confirms this conclusion. We refer the reader to Mr. Raspopin's article for detailed data and give here only his main conclusion. "The interdependence of the condition of stock raising and dairy farming, on the one hand, and the number of dilapidated estates and the intensity of farming, on the other, is beyond question. The uyezds (of Moscow Gubernia) where dairy cattle raising, dairy farming, is most developed show the smallest percentage of dilapidated farms and the highest percentage of estates with highly developed field cultivation. Throughout Moscow Gubernia ploughland is being reduced and turned into meadow and pastureland, while grain rotations are yielding place to multi-field herbage rotations. Fodder grasses and dairy cattle, and not grain, are now predominant... not only on the farming estates in Moscow Gubernia but throughout the Moscow industrial district".

SCALE OF BUTTER PRODUCTION AND CHEESE MAKING

The scale of butter production and cheese making is particularly important precisely because it testifies to a complete revolution in agriculture, which becomes entrepreneur farming and breaks with routine. Capitalism subordinates to itself one of the products of agriculture, and all other aspects of farming are fitted in with this principal product. The keeping of dairy cattle calls forth the cultivation of grasses, the change-over from the three-field system to multi-field systems, etc. The waste products of cheese making go to fatten cattle for the market. Not only milk processing, but the whole of agriculture becomes a commercial enterprise. The influence of cheese production and butter making is not confined to the farms on which they are carried on, since milk is often bought up from the surrounding peasants and landlords. By buying up the milk, capital subordinates to itself the small agriculturists too, particularly with the organisation of the so-called "amalgamated dairies," the spread of which was noted in the 70s.

These are establishments organised in big towns, or in their vicinity, which process very large quantities of milk brought in by rail. As soon as the milk arrives the cream is skimmed and sold fresh, while the skimmed milk is sold at a low price to poorer purchasers. To ensure that they get produce of a certain quality, these establishments sometimes conclude contracts with the suppliers, obliging them to adhere to certain rules in feeding their cows. One can easily see how great is the significance of large establishments of this kind: on the one hand they capture the public market (the sale of skimmed milk to the poorer town-dwellers), and on the other hand they enormously expand the market for the rural entrepreneurs. The latter are given a tremendous impetus to expand and improve commercial farming. Large-scale industry brings them into line, as it were, by demanding produce of a definite quality and forcing out of the market (or placing at the mercy of the usurers) the small producer who falls below the "normal" standard. There should also operate in the same direction the grading of milk as to quality (fat content, for example), on which technicians are so busily engaged, inventing all sorts of lacto-densimeters, etc., and of which the experts are so heartily in favour.

Commercial Grain Farming

In this respect the role of the amalgamated dairies in the development of capitalism is quite analogous to that of elevators in commercial grain farming. By sorting grain as to quality the elevators turn it into a product that is not individual but generic (*res fungibilis*, as the lawyers say), *i.e.*, for the first time they adapt it fully to exchange. Thus, the elevators give a powerful impetus to commodity-grain production and spur on its technical development by also introducing grading for quality. Such a system strikes a double blow at the small producer. Firstly, it sets up as a standard, legalises, the higher-quality grain of the big crop sowers and thereby greatly depreciates the inferior grain of the peasant poor. Secondly, by organising the grading and storing of grain on the lines of large-scale capitalist industry, it reduces the big sowers' expenses on this item and facilitates and simplifies the sale of grain for them, thereby placing the small producer, with his patriarchal and primitive methods of selling from the cart in the market, totally at the mercy of the kulaks and the usurers. Hence, the rapid development of elevator construction in recent years means as big a victory for capital and degradation of the small commodity-producer in the grain business as does the appearance and development of capitalist "amalgamated dairies."

From the foregoing material it is clear that the development of commercial stock farming *creates* a home market, firstly, for means of production—milk-processing equipment, premises, cattle sheds, improved agricultural implements required for the change-over from the routine three-field system to multi-field crop rotations, etc.; and secondly, for labour-power. Stock farming placed on an industrial footing requires a far larger number of workers than the old stock farming "for manure." The dairy farming area—the industrial and north-western gubernias—does really attract masses of agricultural labourers. Very many people go to seek agricultural work in the Moscow, St. Petersburg, Yaroslavl and Vladimir gubernias; fewer, but nevertheless a considerable number, go to the Novgorod, Nizhni Novgorod and other non-black-earth gubernias. According to correspondents of the Department of Agriculture in the Moscow and other gubernias private-landowner farming is actually conducted in the main by labourers from other areas. This paradox— the migration of agricultural workers from the agricultural gubernias (they come mostly from the central black-earth gubernias and partly from the northern) to the industrial gubernias to do agricultural jobs in place of industrial workers who abandon the area en masse—is an extremely characteristic phenomenon.

It proves more convincingly than do any calculations or arguments that the standard of living and the conditions of the working people in the central black earth gubernias, the least capitalist ones, are incomparably lower and worse than in the industrial gubernias, the most capitalist ones; it proves that in Russia, too, the following has become a universal fact, namely, the phenomenon characteristic of all capitalist countries, that the conditions of the workers in industry are better than those of the workers in agriculture (because in agriculture oppression by capitalism is supplemented by the oppression of pre-capitalist forms of exploitation). That explains the flight from agriculture to industry, whereas not only is there no flow from the industrial gubernias towards agriculture (for example, there is no migration from these gubernias at all), but there is even a tendency to look down upon the "raw" rural workers, who are called "cowherds" (Yaroslavl Gubernia), "cossacks" (Vladimir Gubernia) and "land labourers" (Moscow Gubernia).

It is important also to note that cattle herding requires a larger number of workers in winter than in summer. For that reason, and also because of the development of agricultural processing trades, the demand for labour in the area described not only grows, but is *more evenly distributed over the whole year* and over a period of years. The most reliable material for judging this interesting fact is the data on wages, if taken for a number of years. We give these data, confining ourselves to the groups of Great-Russian and Little-Russian gubernias. We omit the western gubernias, owing to their specific social conditions and artificial congestion of population (the Jewish pale of settlement), and quote the Baltic gubernias only to illustrate the relations that arise where capitalism is most highly developed.

Let us examine this table, in which the three principal columns are printed in italics. The first column shows the proportion of summer to yearly pay. The *lower* this proportion is, and the nearer the summer pay approximates to half the yearly pay, the more evenly is the demand for labour spread over the entire year and *the less the winter unemployment*. The least favourably placed in this respect are the central black-earth gubernias—the area where labour-service prevails and where capitalism is poorly developed. In the industrial gubernias, in the dairy-farming area the demand for labour is higher and winter unemployment is less. Over a period of years, too, the pay is most stable here, as may be seen from the second column, which shows the difference between the lowest and the highest pay in the harvest season. Lastly, the difference between the pay in the sowing season and the pay in the harvest season is also least in the non black-earth belt, *i.e.*, the demand for workers is more evenly distributed over the spring and summer. In all respects mentioned the Baltic gubernias stand even higher than the non-black-earth gubernias, while the steppe gubernias, with their immigrant workers and with harvest fluctuations of the greatest intensity, are marked by the greatest instability of wages. Thus, the data on wages testify that agricultural capitalism in the area described not only creates a demand for wage-labour, but also distributes this demand more evenly over the whole year.

Lastly, reference must be made to one more type of dependence of the small agriculturist in the area described upon the big farmer. This is the replenishment of landlords' herds by the purchase of cattle from peasants. The landlords find it more profitable to buy cattle from peasants driven by need to sell "at a loss" than to breed cattle themselves—just as our buyers-up in so-called handicraft industry often prefer to buy finished articles from the handicraftsmen at a ruinously cheap price rather than manufacture them in their own workshops.

This fact, which testifies to the extreme degradation of the small producer, and to his being able to keep going in modern society only by endlessly reducing his requirements, is turned by Mr. V. V. into an argument in favour of small "people's" production!... "We are entitled to draw the conclusion that our big farmers... do not display a sufficient degree of independence.... The peasant, however... reveals greater ability to effect real farming improvements". This lack of independence is expressed in the fact that "our dairy farmers... buy up the peasants' (cows) at a price rarely amounting to half the cost of raising them—usually at not more than a third, and often even a quarter of this cost". The merchant's capital of the stock farmers has made the small peasants completely dependent, it has turned them into its cowherds, who breed cattle for a mere song, and has turned their wives into its milkmaids. One would think that the conclusion to be drawn from this is that there is no sense in retarding the transformation of merchant's capital into industrial capital, no sense in supporting small production, which leads to forcing down the producer's standard of living below that of the farm labourer. But Mr. V. V. thinks otherwise. He is delighted with the "zeal" of the peasant in tending his cattle; he is delighted with the "good results from livestock farming" obtained by the peasant woman who "spends all her life with her cow and sheep". What a blessing, to be sure! To "spend all her life with her cow" (the milk of which goes to the improved cream separator), and as a reward for this life, to receive "one-fourth of the cost" of tending this cow! Now really, how after that can one fail to declare in favour of "small people's production"!

COSMETIC THE PRODUCTIVITY OF SMALL FARMING SYSTEMS

Despite the evidence of the resiliency and productivity advantages of small-scale and traditional farming systems, many scientists and growth specialists and organizations argue that the performance of subsistence agriculture is unsatisfactory, and that agrochemical and transgenic intensification of production is essential for the transition from subsistence to commercial production. Although such intensification approaches have met with much failure, research indicates that traditional crop and animal combinations can often be adapted to increase productivity. This is the case when ecological principles are used in the redesign of small farms, enhancing the habitat so that it promotes healthy plant growth, stresses pests, and encourages beneficial organisms while using labour and local resources more efficiently.

Several reviews have amply documented that small farmers can produce much of the needed food for rural and neighbouring urban communities in the midst of climate change and burgeoning energy costs. The evidence is conclusive: new agroecological approaches and technologies spearheaded by farmers, NGOs, and some local governments around the world are already making a sufficient contribution to food security at the household, national, and regional levels. A variety of agroecological and participatory approaches in many countries show very positive outcomes even under adverse environmental conditions. Potentials include: raising cereal yields from 50 to 200 percent, increasing stability of production through diversification, improving diets and income, and contributing to national food security (and even to exports) and conservation of the natural resource base and biodiversity. This evidence has been reinforced by a recent report of the United Nations Conference on Trade and Growth stating that organic agriculture could boost African food security. Based on an analysis of 114 cases in Africa, the report revealed that a conversion of farms to organic or near-organic production methods increased agricultural productivity by 116 percent.

Moreover, a shift towards organic production systems has enduring impact, as it builds up levels of natural, human, social, financial, and physical capital in farming communities. The International Assessment of Agricultural Knowledge, Science and Technology (AKST) commissioned by World Bank and the Food and Agriculture Organization (FAO) of the United Nations recommended that an increase and strengthening of AKST towards agroecological sciences will contribute to addressing environmental issues while maintaining and increasing productivity. The assessment also stresses that traditional and local knowledge systems enhance agricultural soil quality and biodiversity as well as nutrient, pest, and water management, and the capacity to respond to environmental stresses such as climate.

Whether the potential and spread of agroecological innovations is realized depends on several factors and major changes in policies, institutions, and research and growth approaches. Proposed agroecological strategies need to target the poor deliberately, and not only aim at increasing production and conserving natural resources. But they must also create employment and provide access to local inputs and local markets. Any serious attempt at developing sustainable agricultural technologies must bring to bear local knowledge and skills on the research process. Particular emphasis must be given to involving farmers directly in the formulation of the research agenda and on their active participation in the process of technological innovation and dissemination through Campesino a Campesino models that focus on sharing experiences, intensification local research, and problem-solving capacities. The agroecological process requires participation and enhancement of the farmer's ecological literacy about their farms and resources, laying the foundation for empowerment and continuous innovation by rural communities. Equitable market opportunities must also be developed, emphasizing local commercialization and distribution schemes, fair prices, and other mechanisms that link farmers more directly and with greater solidarity to the rest of the population. The ultimate challenge is to increase investment and research in agroecology and scale up projects that have already proven successful to thousands farmers. This will generate a meaningful impact on the income, food security, and environmental well-being of all the population, especially small farmers who have been adversely impacted by conventional modern agricultural policy, technology, and the penetration of multinational agribusiness deep into the third world.

RURAL SOCIAL MOVEMENTS, AGROECOLOGY, AND FOOD SOVEREIGNTY

The growth of sustainable agriculture will require significant structural changes, in addition to technological innovation, farmer-to-farmer networks, and farmer-to-consumer solidarity. The required change is impossible without social movements that create political will among decision-makers to dismantle and transform the institutions and regulations that presently hold back sustainable agricultural growth. A more radical transformation of agriculture is needed, one guided by the notion that ecological change in agriculture cannot be promoted without comparable changes in the social, political, cultural, and economic arenas that help determine agriculture. The organized peasant and indigenous-based agrarian movements—such as the international peasant movement La Vía Campesina and Brazil's Landless Peasant Movement (MST)—have long argued that farmers need land to produce food for their own communities and for their country. For this reason they have advocated for genuine agrarian reforms to access and control land, water, and biodiversity that are of central importance for communities in order to meet growing food demands.

Vía Campesina believes that in order to protect livelihoods, jobs, people's food security, and health as well as the environment, food production has to remain in the hands of small-scale sustainable farmers and cannot be left under the control of large agribusiness companies or supermarket chains. Only by changing the export-led, free-trade based, industrial agriculture model of large farms can the downward spiral of poverty, low wages, rural-urban migration, hunger, and environmental degradation be halted. Social rural movements embrace the concept of food sovereignty as an alternative to the neoliberal approach that puts its faith in an inequitable international trade to solve the world's food problem. Instead, it focuses on local autonomy, local markets, local production-consumption cycles, energy and technological sovereignty, and farmer-to-farmer networks.

"Greening" the Green Revolution will not be sufficient to reduce hunger and poverty and conserve biodiversity. If the root causes of hunger, poverty, and inequity are not confronted head-on, tensions between socially equitable growth and ecologically sound conservation are bound to accentuate. Organic farming systems that do not challenge the monoculture nature of plantations and rely on external inputs as well as foreign and expensive certification seals, or fair-trade systems destined only for agro-export, offer very little to small farmers that become dependent on external inputs and foreign and volatile markets. By keeping farmers dependent on an input substitution approach to organic agriculture, fine-tuning of input use does little to move farmers toward the productive redesign of agricultural ecosystems that would move them away from dependence on external inputs. Niche markets for the rich in the North exhibit the same problems of any agro-export scheme that does not prioritize food sovereignty, perpetuating dependence and hunger. Rural social movements understand that dismantling the industrial agrifood complex and restoring local food systems must be accompanied by the construction of agroecological alternatives that suit the needs of small-scale producers and the low-income non-farming population, and that oppose corporate control over production and consumption. Given the urgency of the problems affecting agriculture, coalitions that can rapidly foster sustainable agriculture among farmers, civil society organizations (including consumers), as well as relevant and committed research organizations are needed. Moving toward a more socially just, economically viable, and environmentally sound agriculture will be the result of the coordinated action of emerging social movements in the rural sector in alliance with civil society organizations that are committed to supporting the goals of these farmers movements. As a result of constant political pressure from organized farmers and others, politicians will, it is hoped, become more responsive to developing policies that will enhance food sovereignty, preserve the natural resource base, and ensure social equity and economic agricultural viability.

4

Agricultural Marketing Management

INTRODUCTION

Agricultural marketing involves in its simplest form the buying and selling of agricultural produce. In olden days when the village economy was more or less self-sufficient the marketing of agricultural produce presented no difficulty as the farmer sold his produce to the consumer on a cash or barter basis.

TRAINING OF MARKETING PERSONNEL IN AGRICULTURE

Several training courses, viz.:

- 11 months Diploma Course in Agricultural Marketing in Nagpur,
- 4 months Market Secretaries Course at Chandigarh, Lucknow and Hyderabad,
- 6 months Diploma in Livestock Marketing at Nagpur,
- 3 months training for grading supervisors at Nagpur
- 3 months Graders Course at Hubli, Lucknow, and Chandigarh,
- 4 months Training Course in Cotton Classing Centre at Surat,
- 6 months Training Course in Tobacco Grading at Guntur,
- 3 weeks course at Demonstration cum training centre in animal casing at Bombay and New Delhi are being run by the directorate of marketing and inspection.

In addition short term condensed training courses of one week duration organized for the state Government personnel in Kapas grading. As many as 407 candidates in the 11 months Diploma course, 1596 in 4 mths course; 212 in 3 mths course; and 76 in 3 mths grading course were trained by the end of 1972-73.

Market Extension

The directorate of marketing and inspection has set up a separate extension wing for the dissemination of information valuable to producers as well as to consumers. The primary aim of the scheme is to enlighten the producer seller on consumer preference and to advise him on the proper methods of preparation for marketing, grading, storing, packaging, handling and transporting and to improve the quality of the produce and to secure a better return to the growers.

Since 1969, the Directorate of marketing and inspection has in collaboration with the directorate of extension launched a series of 'Agmark' exhibitions known as 'vital link between farm and home'. These have been held in New Delhi, Lucknow, Chandigarh, Trivandrum, Bombay, Calcutta, Bangalore, Nagpur, Ahmedabad, Madras and Hyderabad. These have designed to educate the general public in the field of agricultural marketing to promote quality consciousness among producers.

Marketing Intelligence

With a view to disseminating the marketing intelligence to the interested parties the directorate of marketing and inspection publishes the following journals:

1. Agricultural Marketing—a quarterly journal
2. Marketing Newsletter—a monthly letter
3. Agmark Statistics—yearly
4. Commodity Intelligence Bulletins for tobacco, wool, bristles, potatoes etc.

MARKETING RESEARCH AND INVESTIGATION

The role of market research in the establishment of an efficient system of marketing cannot be overemphasized. In order to be able to introduce reforms one should know the defects and shortcomings of the prevailing system. This calls for and justifies the necessity of intensive research and investigation. Under the third five-year plan an elaborate scheme for conducting survey, and research in the various facets of agricultural marketing has been undertaken by the Directorate of marketing and inspection. Particular attention is being paid to the collection of market information, with regard to price spreads, shifts in marketing practices, consumption pattern, consumer preferences, directional movements, packaging, assembling, transportation, distribution etc. Authentic statistics and data are being collected so that up-to date information may be maintained in respect of all important agricultural commodities. Besides the market surveys cover studies on the organizational aspect of the marketing system, problem oriented studies etc. Under the fourth five year plan a specialized study to estimate the post harvest losses and marketable surpluses of agricultural produce have been taken up on an all-India level.

The Research Wing of the Directorate of marketing and inspection has been further strengthened by creating the Market Research and Planning Cell. The cell has a big component of exports in the field of agricultural marketing research and is expected to increase the tempo of development in this field.

Marketing is a multistage process. For the improvement and development of marketing structure, a co-ordinate approach aiming at removing all the weak links of the marketing chain is essential. A package of improved marketing services in the form of regulated markets, grading, weighing, storing, transporting, handling services and marketing finance need to be made available to ensure the producer a fair return from his production efforts and a better share in the price paid by the consumer. At the same time, market research programmes should be oriented to the developing of an orderly and efficient marketing system. This is a crucial time in the development of agricultural marketing when the country is poised to enter an era of production surpluses. A piecemeal approach at this stage can be disastrous and can nullify the advantages gained by the farmer on the production front.

MARKETING FUNCTIONS OF AGRICULTURE

In modern marketing the agricultural produce has to undergo a series of transfers or exchanges from one hand to another before it finally reaches the consumer.

This is achieved through three marketing functions:

1. Assembling,
2. Preparation for consumption
3. Distribution.

Concentration pertains to the operations concerned with the assembly and transport from the field to a common assembly field or market. The produce may be taken direct to the market or it may be stored on the farm or in the village for varying periods before its transport. It may be sold as obtained from the field or may be cleaned, graded, processed and packed either by the farmer or village merchant before it is taken to the market. Some of the processing is done not because consumers desire it, but because it is necessary for the conservation of quality. At the market the produce may be sold by the farmer direct to the consumer or more usually through a commission agent or a broker. It may also be purchased by traders, wholesalers or retailers. The transactions may be carried out by direct negotiation or through middlemen, by barter or cash, by open or under cover auction, on the spot or in future markets. The transactions take place at one or more levels in the primary, secondary or terminal markets or all three. Distribution (dispersion) involves the operations of wholesaling and retailing at various points. By a series of indispensable adjustments and equalising functions, it is the task of the distribution system to match the available supplies with the existing demand.

MARKETING FUNCTIONARIES (AGENCIES)

The transfer of produce or goods takes place through a chain of middlemen or agencies. In the primary market the main functionaries are the producer, the village or itinerary merchant, pre-harvest contractors, commission agents, transport agents etc. In the secondary market the processing and manufacturing agents are the additional functionaries. Financing agents such as shroffs, banks and co-operatives may also take part. In the terminal or export market the commercial analyst and shipping agent also gets involved in the transfer of goods.

The functionaries have their own set up. They may be individuals, partners or co-operatives who may buy and sell on ready and future basis at a price determined by forces of supply and demand. Each functionary renders some service in the process of marketing and also earns a varying margin of profit for himself. This procedure makes marketing rather complicated and inflates the price of the produce. The nature of some of the agricultural products for example their bulk form and perishability and their seasonal availability further add to the complexity of agricultural marketing.

MARKETING IMPROVEMENTS

India being a primary producing country, agriculture plays a vital role, both as an essential infrastructure and a development component in generating and sustaining a higher national income. Out of a national income of about ₹38,921 crores in 1972-73 as much as ₹17,500 crores or about 44.9 per cent is contributed by agriculture and allied sectors. It is estimated that about 50 per cent of the agricultural produce is available as marketable surplus. The marketing system in India provides sustenance for about 3 million persons who are engaged in performing various marketing function. In the field of exports too, the agricultural sector accounts for about 50 per cent of the total value.

The process involved in the disposal of such a substantial produce of great economic importance are significant not only for the farmer but also for the country as a whole. The unreasonably low return that the farmer gets for his produce and the excessive margin of profit retained by the intermediaries attracted the Government's attention and it was felt that the economic condition of the agriculturists could not be improved unless determined steps were taken to establish an orderly system of marketing in the country.

GOVERNMENT REGULATORY PROGRAMMES

With this object in view, a number of marketing surveys were conducted by the Directorate of Marketing and Inspection which revealed the shortcomings in the country's marketing system. A rectification of these deficiencies was sought to be achieved by rationalizing various activities and standardizing various practices in the markets through legislation or otherwise. The primary objective of improving the system of agricultural marketing was not only to remove the handicaps from which the producer-seller was suffering but also to increase his income by ensuring him a fair price.

REGULATION OF MARKET

Prevailing market practices and market charges made a deep cut in the share of the producer in the price paid by the consumer. Some of the market charges were authorized whereas others were more than what the service rendered warranted. It was felt that a remunerative price to the producer could only be ensured if the market practices and market charges were regulated and rationalised. And thus the regulation of the markets has been given a high priority in the various Five Year Plans. The markets are sought to be regulated through an Act of each legislature. The Act is generally known as the Agricultural Produce Markets Act and it is provided for the removal of various malpractices widely prevalent in the markets for the settlement of disputes between sellers and buyers and for the promoting of orderly marketing of farm produce in general. Various state Governments have made considerable progress in this field by bringing in the necessary legislation. The Acts enabling the respective states to regulate the markets generally provide for the notification of market areas and the commodities to be covered in the act in different areas. A 'Marketing Committee' consisting of the representatives of growers, traders, merchants, local bodies and Government nominees administers the working of each market. The functions of the market committee are to frame bye-laws, define local market prices, fix market prices payable to various functionaries, license the functionaries, settle disputes, supervise weightment and promote the development of orderly marketing in general. The committee is generally empowered to raise funds for its working by levying a small fee on the produce bought and sold in the market in addition to the license fees received from the functionaries.

Market Surveys

A survey conducted in 496 markets in 1961-62 has shown that after the regulation the market charges have been reduced by 48 per cent. Another survey of selected commodities revealed that for some commodities market charges have been reduced by as much as 98 per cent.

Although the first market to be regulated was Karanja in 1886 the regulation of markets did not make headway till the first Five Year plan. It got a fill up in the second and third five year plan.

The states and union territories which have regulated the markets are: Andhra Pradesh(335), Bihar(62), Chandigarh(1), Delhi(3), Gujarat(212), Mysore(102), Madhya Pradesh(233), Haryana(80), Kerala(6), Maharashtra(212), Orissa(34), Punjab(95), Rajasthan(90), Tamil Nadu(136), Tripura(1), Uttar Pradesh(264), Goa, Daman and Diu(1), West Bengal(15), and Himachal Pradesh(5). The states and union territories which are yet to regulate markets are Assam, Andaman and Nicobar islands, Arunachal Pradesh, Dadra and Nagar Haveli, Jammu and Kashmir, Laccadive and Minicoy islands, Meghalaya, Mizoram, Nagaland, and Pondicherry. Necessary measures to regulate markets in these states and union territories are at various stages of progress. It is expected that these states and union territories will regulate the markets by the end of the Fifth plan.

All the states where necessary legislation has since been passed, have formulated phased programmes for the regulation of markets. By the end of the fifth plan it is expected that all the wholesale assembling markets would be brought within the regulatory orbit.

As a result of this scheme, excessive commissions and other market charges have been substantially rationalized. Unauthorized and arbitrary deductions have been prohibited and malpractices stopped. The issue of sale slips by licensed commission agents to the sellers, indicating the details of sale proceeds, deductions effected etc. has been made obligatory. Weightment is also done by licensed weightmen of the market committee.

The dissemination of marketing information and news is one of the functions of the market committees. This is done through the displaying of the prices prevailing in the market and also in the neighbouring markets on the notice boards and announcements through loud speakers at regular intervals.

This information is also supplied to the Central Government and the State Government and also to other market committees. Arrangements have been made for the maintenance of reliable statistics arrivals, sales, stocks, prices etc. which are maintained by the market committees.

It is also obligatory on the part of the market committee to provide the market yards with the necessary amenities. In some of the markets that have been regulated amenities like rest houses, cattle sheds and water troughs have been provided by the market committees for the convenience of the producer sellers. Facilities for grading before sale and storage have also been provided.

A survey of 500 regulated markets was undertaken by the Directorate of Marketing and Inspection in the Ministry of Agriculture in 1970-71 and 1971-72, with a view to assessing the adequacy and efficiency of the existing regulated markets and highlighting their drawbacks and deficiencies and suggesting measures to develop them. One of the most important drawbacks has been the inadequate financial resources of some of the market committees. During the fourth plan, a central sector scheme was drawn up by the Ministry of Agriculture to provide a grant at 20 per cent of the cost of development of market, subject to a maximum of ₹ 2 lakhs. The balance will have to be provided by the commercial banks.

An important development in the field of regulated markets is the keen interest taken by the International Development Agency (IDA) in the development of the infrastructure in regulated markets. The IDA is financing the development of infrastructure in 50 markets of Bihar. The World Bank has approved a loan assistance of 6.5 crores to Karnataka also for the development of markets.

CONTRACT TERMS

Under the existing trade practices, the sale of produce in a primary market takes place on the basis of the visual inspection of the goods, and in the secondary and terminal markets on the inspection of the samples. Thereafter the buyer and the seller decide upon the terms either orally or through written contracts. The contract terms specify the quality and quantity of the produce, the time and place of delivery, the price and terms of payment, handling and incidental charges, the procedure for settlement of disputes and penalties. The terms of contract were not standardized and thus varied for every individual transaction, and were more favourable to the buyer.

With a view to improving trade practices, All-India standard contract terms have been drawn up for a number of commodities. In standard contract terms the definition of quality and allowances in respect of refraction, damaged goods have been specifically standardized though the adoption of these standard contract terms by traders is voluntary they have to a large extent strengthened the position of the producer-seller and have improved the quality of the product marketed.

STANDARDIZATION AND GRADING

In order to gain the confidence and establish a rational relationship between the quality of a produce and its price, it is necessary to devote some attention to the proper preparation sifting and sorting of a material according to certain attributes before it is taken to the market. This is sought to be achieved by grading the produce in conformity with certain accepted quality standards *viz.* shape, size, form, weight, and other physical and technical characteristics. The produce brought to the market is very often contaminated with dust, stones and other foreign matter added either deliberately or by accident. Sometimes the produce is immature or not properly dried or contains shrivelled grains or damaged and rotten material. Such a produce brings a lower price to the farmers. Care should be exercised while assembling the produce of different farmers so that the good material is not mixed with the inferior material brought in by some farmers.

The Government of India had recognized the need to introduce the standardization of agricultural produce which would enable the farmers to derive the benefits of grading in terms of fair practices according to the prescribed standards and enacted the Agricultural Produce Grading and Marking Act in 1937. The Act empowers the central Government to prescribe grade standards indicating the quality of articles included in the schedule and specify grade designation marks to represent particular grades or qualities. The Act provides for the grading and marketing of agricultural produce. The grade standards prescribed under this act are based on both physical and chemical characteristics and are formulated after analysing representative samples of each commodity collected from different regions and different seasons. Besides the international standards and special requirements of overseas consumers are also taken into account while formulating these standards for the commodities which are exported. The grade standards are reviewed and amended from time to time in the light of the shift of the pattern of production and trade and changes in the consumer's preferences. The grades are designated as the 'Agmark' grades.

A central Agmark Laboratory at Nagpur with sixteen regional laboratories at Guntur, Madras, Bombay, Kanpur, Cochin, Rajkot, Calcutta, Sahidabad, Jamnagar, Bangalore, Patna, Tuticorin, Virudhunagar, Mangalore, Alleppey and Kozhikode are assisting to provide adequate laboratory facilities for fixing grade standards for new commodities, for revising old grade standards and for routine quality control work.

Grade for Export

Grading of agricultural produce under the A.P. (G and M) Act is voluntary. Exports of certain agricultural commodities have however been prohibited unless duly graded and marked in accordance with the grade standards laid down under the A.P. Act 1937. The power to so prohibit exports was derived under the provisions of the sea-customs act.

The Directorate of Marketing and Inspection under the Ministry of Agriculture exercises a three tier control on the quality of agricultural commodities that are graded under 'Agmark' before they are exported. This is done through inspection.

Grading for Internal Trade

Commodities such as cotton, ghee, butter, rice, wheat, atta, gur, eggs, arecanut, potatoes, fruits, bura, pulses, vegetable oils and ground spices are being presently graded under 'Agmark' on voluntary basis.

Grading at Farmers Level

The grading of agricultural commodities under 'Agmark' has been consumer oriented. Generally the grading was done at the level of the traders. At this stage the producer was not a direct beneficiary of the grading scheme. It was felt the need to introduce grading at the producers' level. Thus the Directorate of marketing and inspection introduced a scheme for setting up commercial grading units.

Grading of Fruits and Vegetable Products

With a view to exercising quality control over fruits and vegetables the Government promulgated the Fruits Product Order under the essential Commodities Act. The preservatives and colors to be used are also clearly laid down. The order also stipulates the hygienic and sanitary methods which must be adopted by the manufacturers. The license for all this is issued by the executive Director, food and Nutrition board.

The total number of licensed factories was 1194. The total average production of fruits and vegetable products in the country during 1970 has been estimated as 156 thousand tonnes valued at ₹32.66 crores. Some of the products are very popular in foreign markets and are a good source of foreign exchange. In 1970 alone fruit products worth ₹2.92 crores were exported from India.

Regulation of Cold Stores

Most of the problems relating to the marketing of fruits and vegetables can be traced to their perishability. Perishability is responsible for high marketing costs, market gluts, price fluctuations and other similar problems. At low temperature, perishability is considerably reduced and the shelf life is increased and thus the importance of cold storage or refrigeration.

The first cold store in India was reported to have been established in Calcutta in 1892. However significant progress in the expansion of the cold storage industry in the country has been made only after independence.

An ad-hoc survey of the cold stores carried out by the directorate of marketing and inspection in 1955 showed that the total available cold storage in the country was only 77 thousand tonnes. The survey also highlighted the need to regulate the cold storage industry in a planned manner.

With a view to ensuring the observance of proper conditions in the cold stores and to providing for development of the industry in a scientific manner, the Government of India and the ministry of agriculture promulgated an order known as "cold storage order, 1964" under Section 3 of the Essential Commodities Act, 1955. The order is applicable in respect of every cold store with a capacity more than 8.4 cu m. The jurisdiction of this Order extends to the whole of India except West Bengal. Under this order, it is obligatory on every operator of the cold store to obtain a license from the Licensing Officer before using the installation for storing any food stuffs *e.g.* fruit, vegetables, meat, fish, dairy products. The Agricultural Marketing Advisor to the Government of India is the Licensing Officer.

The directorate of marketing and inspection is enforcing the cold storage order. The field staff posted in the regional and sub offices located in different states regularly inspects the cold stores and offers necessary guidance for better and scientific preservation of foodstuffs. Besides the directorate of marketing and inspection gives general guidance on all technical matters concerning the setting up of cold stores to the intending entrepreneurs.

The Government of India constituted a Central Cold Storage Advisory Committee consisting of official and non official members, representing the growers, owners, machinery manufacturers, research organizations etc. The Committee advises the Government on all matters pertaining to the enforcement of Cold Storage Order and the future development of the industry. At the end of 1973-74 there were 1503 cold stores in the country with a capacity of 18.70 lakh tonnes. Uttar Pradesh had the maximum number of cold stores of 9.37 lakh tonnes followed by Bihar with 169 cold stores with 2.10 lakh tonnes capacity and West Bengal with 133 cold stores with 3.15 lakh tonnes capacity. Out of 1217 cold stores licensed during 1970, 1021 representing 84 per cent of the total were owned by the private sector, whereas 121 and 75 were owned by the public and co-operative sector respectively. The respective capacity is 13.88 lakh tonnes, 0.25 lakh tonnes and 0.73 lakh tonnes. Though numerically the cold stores in the co-operative sector were less than those in the public sector the capacity was more. The National Co-operative Development Corporation (NCDC) has formulated a scheme for financial assistance for setting up new cold stores in the co-operative sector.

Potato is the most important commodity which is presently placed in the cold stores accounting for as much as 92 per cent of the total capacity in the country. The remaining 8 per cent is being utilized for other perishables *viz.* fruits, vegetables, meat, sea-foods, dairy and poultry products.

With the attainment of self-sufficiency in the production of food grains, greater attention is being paid to the increased production of protective foods such as fruits, vegetables, fish, poultry and dairy products. These products being highly perishable are required to be kept in cold stores. Owing to the inadequate cold storage facilities at present considerable losses occur in the case of these commodities. There is enough scope for expanding the cold-storage industry with a view to providing facilities for preserving and prolonging the shelf life of these protective foods which are essential for human health.

Consumer Protection

In the marketing process the producers and consumers are the two weak ends of the chain. It is incumbent on the part of the Government to protect the interests of both of them. Producers are sought to be protected through the regulation of the markets, grading at the producers level and other similar measures. Consumer's interests on the other hand are safeguarded by grading under 'Agmark' at the level of traders. The progress towards making the consumer quality conscious is slow. With the growing popularity of semi-processed foods the danger of sub-standard food articles being marketed has become manifold. With this in view steps have been taken in many directions, *e.g.* Acts relating to grading and standardization of agricultural commodities, certification marks in respect of manufactured goods, pure food laws, laws relating to weights and measures, and laws relating to the manufacture of fruits and vegetable products have been passed and enforced. Commodities such as ghee, vegetable oils, butter, honey and powdered spices are being graded and marked under 'Agmark' under the provisions of the Agricultural Produce (Grading and Marking) Act 1937. The Agmark attempts to provide a third party guarantee for the consumer. It not only certifies the purity of the product but also gives an indication of their quality by the grade-designating mark. The economic incentive to the producer-manufacturer is reflected in the premium the Agmarked product fetches in the market over the ungraded products.

Running parallel to the enforcement of laws, certain measures to complement the effort of achieving the overall objective of consumer protection are also being adopted. Monopoly procurement of food grains being operated by the Food corporation of India and other state departments achieves the two objectives of ensuring a remunerative price to the farmers and a uniform and reasonable price commensurate with the quality of the produce to the consumers.

The task of consumer education and protection is formidable and the Government machinery alone will be inadequate to accomplish this task. There is a need of active consumers' movement as in other countries. The Consumers' Guidance Society has been established with the objective of educating the consumers about their rights and responsibilities and advising them on the quality of products available in the market. Similarly it has advised the producers and manufacturers to abide by such standards as are necessary for the health and safety of the users.

Co-operative Marketing

The existing institutional structure of co-operative marketing is such that the co-operatives are functioning at the primary level, at the secondary level (taluka or district) and at the state level. In pursuance of a recommendation of the Dantwala Committee, efforts have been made to persuade the state Governments to divert the middle tier *viz.* the district marketing federations, of the functions legitimately falling within the purview of the state or primary marketing societies, so that a two-tier system can be brought into operation. The co-operatives in different states have been federated into a central level federation *viz.* the National Agricultural Co-operative Marketing Federation (NAFED). At the end of 1960-70 as many as 3335 primary cooperative marketing societies were operating. Of these more than 500 were specialized commodity marketing societies for special crops, cotton, fruits and vegetables. In certain states intermediate organizations at district and state levels have also been established. At the end of 1960-70, 232 such societies were functioning.

They included some commodity federations also. At the state level 28 apex co-operative marketing federations are functioning. They are normally handling all the commodities. There are a few apex cooperative societies which are handling exclusively a particular commodity *e.g.* two apex societies in Gujarat are handling cotton, one is handling fruits and vegetables and one in Uttar Pradesh is handling sugarcane only. In order to strengthen and develop co-operative marketing to an extent where it may have an impact on the marketing of agricultural produce, several measures have been initiated. Steps are also being taken to see that there is an effective co-ordination between the state cooperative departments and their counterparts dealing with agricultural marketing. Consequently there has been a significant expansion in the operations of marketing co-operatives. This is exclusive of the value of agricultural requisites and consumer articles handled by the marketing cooperatives. The main commodities marketed by the cooperatives were sugarcane, cotton, oilseeds, fruits, vegetables and plantation crops.

In pursuance of the decision of the Government, a scheme of outright purchases of agricultural produce by the co-operative marketing societies was launched in 1964-65 in 200 selected marketing societies. The basic theme of this scheme has been to bring the small producers within the fold of co-operative marketing together for such farmers. To provide the necessary financial backing the scheme envisaged the creation of a price-fluctuation fund. The fund envisages meeting losses if suffered by the co-operative marketing societies at different levels as a result of outright purchases of agricultural produce.

The scheme has been operative in several states. During 1969-70 the value of the agricultural produce purchased under the outright purchase scheme was about 34 crores. As a result of this scheme there has been a greater involvement of cooperative societies in the marketing of agricultural produce.

Besides this scheme has given impetus to inter-state trade by the cooperatives. They transacted about ₹ 66.55 crores worth of agricultural produce during 1969-70.

Interstate Trade

The co-operative marketing societies are devoting increasing attention to interstate trade in agricultural produce. The main commodities are wheat, pulses, plantation crops, copra, spices, fruits and vegetables. The bulk of the transactions were made by the Punjab Apex Federation. During 1970-71 the NAFED acted as the agency of the co-operatives of Jammu and Kashmir for marketing their apples outside the state.

Co-operative Export of Agricultural Produce

The export of agricultural produce by the co-operative sector continues to be a growing activity. The bulk of the exports are made by National Agricultural Cooperative Marketing Federation Ltd., which accounted for exports worth ₹5.64 crores. Besides, the Gujarat State Cooperative Marketing Federation Ltd., the Khanna Cooperative Marketing Federation Ltd., the Kerala State Cooperative Marketing Federation Ltd., the Jalgaon district fruit sales societies and the coconut oil millers society also cooperated to exports to countries like Kuwait, Malaysia, Ceylon, Singapore, Bahrain, Doha, Dubai, Iran, Muscat. The Khanna Solvent Extraction Plant in the Punjab State exported de-oiled cake worth ₹35 lakhs during 1969-70. The main commodities exported by the cooperatives were pulses, chillies, onions, pepper, de-oiled cake, potatoes and kardi extraction meal. The exports were mainly made to Ceylon and other important markets were Mauritius, Kuwait, Doha, Bahrain, Hong

Kong, the USSR, the UK, Iran, Czechoslovakia and France. Some of the traditional items of export have been marketed in non traditional areas. Pulses were exported to Cuba and onions to Malaysia and Singapore. The cooperatives also assisted various agencies to export agricultural commodities. In this connection exports of coffee and raw sugar were made. During 1970-71, NAFED exported agricultural produce worth ₹5.26 crores. The Jalgaon District Fruit Sale Societies Cooperative Marketing Federation Ltd., directly exported bananas worth ₹34 lakhs to Kuwait and Bahrain Islands.

Co-operative Cold Stores

Co-operatives are also to facilitate the storage and marketing of perishable commodities especially seed potatoes. By the end of the Third plan the cooperatives had established 87 cold stores and the target for the fourth plan was set at 45 more. At the end of December 1971 there were as many as 96 cooperative cold stores with a capacity of 1.42 lakhs tonnes.

TYPES OF MARKETING INTERVENTION

There are many different ways in which NGOs or CBOs may intervene to improve access to agricultural markets.

In this section interventions are discussed in eight non-exclusive categories that describe aspects of the intervention strategy:

1. Intended beneficiaries
2. Skills and training
3. Access to agricultural inputs
4. Agro-processing technologies
5. Marketing linkages
6. Credit programmes
7. Marketing information
8. Holistic approaches

Any particular marketing intervention may comprise elements from several categories (*e.g.* inputs and training, or technology, training and finance). The concepts and experiences associated with each category are reviewed, permitting some preliminary conclusions on the more promising strategies.

INTENDED BENEFICIARIES

Different NGO and CBO marketing initiatives operate with different beneficiary or client structures: some work with whole communities, some with groups and others with households/individuals. The choice of appropriate structure will depend on a number of factors. Research carried out by the Plunkett Foundation and experience of CARE's Development through Conservation project in Uganda suggests that working with village associations or whole communities is more difficult than working with smaller groups. The latter are more focused, more specialized and more likely to have a common goal. Many NGOs and CBOs implement their marketing interventions with groups or associations. Not only are there advantages to the NGOs and CBOs of working with groups, but there may be advantages for the farmers themselves of marketing collectively. The groups can be existing groups (*e.g.* women's groups, savings and credit groups, social groups and so on) or newly formed groups.

The potential advantages of farmers collectively addressing marketing constraints include:

- Economies of scale, through joint purchasing of inputs and joint marketing of products;
- Improved access to finance, where credit organizations favour group loans, or where pooled resources provide the necessary down payment; this can overcome problems of larger investment needed in, for example, processing technologies, storage facilities or transport;
- Collective bargaining power;
- Lower transaction costs (for producers and traders).

Yet there is much evidence that, in general, small enterprises owned and managed by individuals are more successful than group enterprises. Technology adoption in particular is felt to be better amongst individuals than groups, although groups can be an appropriate vehicle for technology transfer if the technologies are subsequently employed by individuals. This is echoed by String fellow *et al.* who argue that group enterprises are more likely to succeed when based on joint marketing rather than joint management/ ownership of assets, because the latter requires more complex skills and experience. It has also been found that external organization and management of groups can prevent the development of entrepreneurial skills. There are no definitive rules on which structure is appropriate. It will depend on the type of intervention and the objectives of the individuals and NGOs/CBOs. On the one hand, NGOs need to remain open to working with groups where appropriate, recognizing the potential to build capacity, reduce transaction costs, and introduce activities with a higher investment threshold. However, an understanding of the reasons behind group formation is important and care should be taken not to over-burden groups formed for social rather than economic reasons. Sometimes individuals, often women and particularly in rural areas, prefer to work together. Furthermore, commonly cited problems attributed to groups per se have been found to be less important when there is strong group leadership and cohesion, and when there is good group organization before any external intervention takes place. String fellow *et al.* point out that a non-interventionist approach – letting producers decide for themselves whether they operate as individuals or groups – allows individuals to develop appropriate structures to build necessary skills and solve their own problems. People will not work well within an imposed structure. As Gibson succinctly states: "Enterprises working in a market environment (whether collectively owned or individually owned) have ultimately to make a surplus in order to survive. The 'bottom line' for NGOs is to support the structures which will work best according to this criterion."

Rural Poor

The majority of NGOs and CBOs deliberately target their activities at poor households and poor communities. Those that focus on agricultural production, processing and marketing are often found in remote rural areas. This focus has implications for the approach adopted by the NGOs/ CBOs and the marketing initiatives they develop. Whereas the trend in marketing approach adopted by NGOs has moved towards a more business-oriented, facilitative approach in recent years, some authors argue that this is less likely to succeed in remote rural areas than in less remote, higher-potential areas. Moreover, within rural communities, some poor individuals and groups are considered too remote and disadvantaged to be able to benefit from marketing interventions. The most vulnerable households and women-headed households in rural areas are particularly disadvantaged; they tend to have more restricted access to information and services and tend to be more risk-averse than other households. Some households may only benefit indirectly from a marketing intervention through any impact it has on the labour market. Consideration should also be given to the types of intervention most appropriate to a particularly disadvantaged target group, particularly where the transfer of hardware and entrepreneurial skills are envisaged. For example, food processing technologies are often predominantly adopted by the rural elite, due to their more developed entrepreneurial skills and access to finance. This may nonetheless reduce poverty, through improved markets and employment generation, but the strength of these linkages varies, making it difficult to generalize. Certainly more information is needed on the extent to which the poor benefit directly and indirectly from NGO and CBO marketing interventions, and about which approaches and types of intervention are having most impact.

Women In common with other organizations, NGOs and CBOs often find it very difficult to effectively target benefits to women. Yet women are frequently amongst the poorest members of rural communities and face particular constraints in improving their livelihoods. Moreover, their traditional role in food crop production and (sometimes) domestic marketing, as well as the influence they have over child welfare and nutrition, makes them an obvious target for poverty-focused agricultural marketing interventions. In their review of women's role in post-harvest operations, Gordon *et al*. conclude that:

"Where interventions are intended to benefit the poorest women, attention should be focused on particular issues:

- The needs of female-headed households, which feature disproportionately amongst the poor;

- Crops and processes used in marginal areas;
- Carrying fuel and water, because so many women are affected;
- How poor women earn income – so that new technology really does benefit them;
- Understanding the broader processes which determine how benefits are distributed;
- Household level and informal sector activities, where the poorest people earn their living."

Goodland *et al.* make a number of suggestions on how women's participation in rural finance programmes can be increased, the spirit of which is equally relevant here:

- Out-of-hours opening or mobile services, or locating services in places women frequent, for example, marketplaces;
- Relaxing literacy requirements;
- Flexible collateral requirements, for example, accepting jewellery rather than land;
- Allowing loans of a suitable size (usually small).

SKILLS AND TRAINING

The types of training offered by NGOs in support of marketing initiatives can include group strengthening, general extension, marketing and specialized training.

TechnoServe and Oil Palm Processing in Ghana

TechnoServe is an NGO whose activities focus on the provision of food processing technologies for rural communities. The initial focus of its work was small-scale oil palm processing and extraction in Ghana, but it has subsequently expanded to include processing service centres, inventory credit schemes and production and processing of non-traditional export crops. The purpose of the oil palm scheme was to build on traditional processing (which was laborious and slow) and to exploit the potential for expansion in a strong market, by introducing small-scale mechanized oil mills to community-based groups. The approach adopted by TechnoServe went beyond providing the technology and technical assistance. It was recognized that entrepreneurial skills in the communities were weak, and that group development activities and financial and business training (including linkages to formal credit and extension services) were necessary. This integrated training and support package clearly contributed to the sustainable adoption and management of the oil processing enterprises. However, the Techno Serve approach has also been criticized. The training and support provided to groups are heavily subsidized. Case studies of the oil palm enterprises indicate that the skills are held by only a small number of active group members, and that because of social norms, there is a reluctance to pass these on. A more serious criticism of the approach is that although it has been successful in transferring skills specific to the oil palm enterprise, it has not managed to create more general entrepreneurial skills and characteristics amongst the group members.

Training and Extension

The starting point of a number of CBO/NGO marketing initiatives is production. This is because to market successfully, farmers need to produce and sell what is in demand, at a profit. Often existing markets could be accessible to farmers (either on their own or through linkages with traders), but marketing is constrained by the low volumes or poor quality of farmers' crops. Improved production practices are important for increasing yields of existing crops, new varieties and new crops. Yet government extension services are often lacking or extremely under resourced. NGOs, therefore, often assume a role in providing, or facilitating the provision of, relevant extension information. This may be through strengthening existing extension services or through the establishment of alternative services, such as farmer-extension schemes.

When the CARE Egypt Agricultural Reform Programme began, it focused on improving crop and livestock production. Now, however, the project has become more market oriented and requires farmers to examine the market for their products prior to improving production. Other NGOs and CBOs provide marketing training to groups and individuals. This includes training in production and marketing systems, constraints and opportunities, market demands (products and service) and how to assess whether products can be supplied profitably. The benefits of providing producers with this type of training are that it develops their capacity to analyse markets for themselves and, therefore, allows them to respond to changing market opportunities and threats. They are not dependent on the NGO/CBO to identify marketing opportunities for them in the longer term. It is also important for NGOs and CBOs not to create parallel services and for different development organizations and governmental bodies to co-ordinate the training and services they provide.

In Uganda, for example, a task force was established to prepare guidelines on how to integrate NGO activities into district agricultural programmes. Specialized or vocational training is usually provided to individuals or groups when the marketing intervention relates to a new product or new technology, and requires new skills. Examples include production targeted to a new niche market or particularly quality-conscious export market, or training in the use of an agro-processing technology, such as an oil press. Increasingly organizations which provide skills and training for enterprise development are charging trainees a fee. This helps NGOs and CBOs with limited funding to reach a wider audience. More significantly, it has been found that charging a fee increases the proportion of trainees who actually make effective use of their training. The entrepreneur, making the decision to invest money and time in training, is in effect making a risk assessment.

Group Formation and Strengthening

The advantages of working with groups have already been highlighted. Despite these, group establishment and operation have generally proven more difficult in practice than expected. Some organizations undertake group strengthening and training (*e.g.* in recruitment and management, structure and leadership, financial and business management skills and so on) before marketing activities are initiated, or as part of the marketing intervention. The case of oil palm processing with the NGO TechnoServe is a good example. The danger of not assessing the performance of existing groups and providing any necessary training in group strengthening is that marketing activities implemented with these groups can fail due to general group weaknesses, rather than problems with the marketing activities per se. Stringfellow *et al.* studied farmer co-operative enterprises and their findings highlight the importance of not over-estimating group capacities, and the need for long-term involvement in building group capacities. Often groups fail because they have been formed too quickly and too much is expected of them. They also found that group enterprises are more likely to succeed when based on joint marketing rather than joint management/ ownership of assets which requires more complex skills and experience. It is also important to consider the most appropriate group size and whether there is a culture of working as a group. The Co-operative League of the USA (CLUSA) working with CARE in Mozambique has developed a thorough approach to the development of farmers' associations. They have attempted to make this approach more sustainable by developing the capacity of a local NGO to take-up and continue these activities, and by promoting a structure within the farmers' associations that permits ongoing development. The approach has been criticized for being too costly, but the results to date, albeit over a short duration, are nonetheless very impressive. There is now a need to critically evaluate this approach, to identify the direct and indirect subsidies provided by the NGOs, and assess which components have the most prospects of sustainability.

IMPROVING ACCESS TO AGRICULTURAL INPUTS

Poor access to inputs directly influences the level and quality of production. Even in the poorest parts of Africa, there is still demand for farm implements and good quality seed. Less-poor farmers may make selective use also of fertilizer and pesticides. Input subsidies are a particularly vexed issue. Some argue that they are needed to provide a short-run boost to production and incomes. Yet they are also disruptive and undermining of sustainable commercial development. At a workshop in Uganda, participants highlighted the negative effect of farm input relief programmes in neighbouring countries on the development of commercial input supply networks in Uganda. In Malawi, the starter-pack scheme (distribution of free seed and fertilizer) implemented in 1999 and 2000 has boosted production there, but it has also deprived other low-income farmers in neighbouring countries (notably Mozambique) of a traditional outlet for their surplus production. Gordon discusses five sets of issues affecting access to inputs: affordability; availability; information; risk and uncertainty; and the overall commercial context. Although credit is often assumed to hold the key to improved access, other ways to improve affordability are also identified: timing input sales to coincide with times when farmers have cash; selling inputs (*e.g.* seed) in small pack sizes suited to small producers; and lowering prices, by making cost reductions in distribution and marketing (*e.g.* through bulk purchases, transport sharing arrangements, and farmers' groups taking on more responsibilities).

Whilst NGOs may play a role in providing credit or promoting some of these other strategies, a role for CBOs (*i.e.* farmers associations) is much more apparent. In countries that have succeeded in significantly improving access to inputs, farmers' associations have acted as a key vehicle for input distribution. Many consider the physical availability of inputs to be a more important constraint to access, with thin and unreliable rural distribution networks in most African countries. A recent development involves innovative approaches to the promotion of input stockist networks by NGOs, illustrating what can be achieved through constructive partnerships between the commercial, private non-profit, farming community and government sectors. These initiatives typically involve training stockists and may involve loans or loan guarantees.

There is also a growing interest in ways to improve and build on traditional informal seed systems. Information constraints are also important – be they in terms of information gaps (basic research on fertilizer response, for instance) or information flows. Although farming is inherently risky, better information reduces uncertainty, enabling farmers to make more informed production decisions. Gordon concludes with some general observations, which also have relevance to the activities of NGOs: "In addition to policies aimed towards the general development of rural economies, a number of more specific policy recommendations are made: avoid actions which undermine the development of sustainable commercial input supply networks; support input markets by setting standards and regulations, and providing information and training; promote synergistic partnerships between commercial, private non-profit, farming community and government sectors; and fill critical research and information gaps."

AGRO-PROCESSING TECHNOLOGY

Small-scale farming households in remote rural communities generally find that they operate in markets comprising many producers of undifferentiated products, which leads to price competition and low profit margins. Access to processing technology can provide new market opportunities by reducing perishability or adding value in other ways.

Processing technologies can range in scale from household-level 'lowtech' processing to fully mechanized factories. Household level processing has two main functions. It can add value and it can preserve the product, thereby increasing the time available for marketing. Other advantages of small-scale agro-processing enterprises are that they can create employment at low levels of investment that make effective use of local resources. Enterprises owned and managed by individuals or households are often more successful than group enterprises, so technology development organizations need to be aware of this need for small-scale technologies.

However, many NGOs and CBOs work with farmer groups either because it is the structure that the farmers prefer (there may be a culture of group activity) or because the NGO/CBO prefers this approach (due to financial and coverage reasons). Yet group approaches to the adoption of agroprocessing technology are often weak and entrepreneurial skills are less evident than when working with active individuals. In these circumstances, careful group selection is required, as well as consideration of ways in which the need for entrepreneurial skills can be reduced by introducing a third party, such as a private company offering marketing services to small-scale processors. It is necessary to ensure that there is market demand for the technologies and/or their end-products and that the technologies are appropriate (*e.g.* taking into account gender issues), rather than developing technologies for their own sake. It is also important that the wider enterprise environment is considered and that institutional arrangements enable smallholder access to markets. Also, as will be examined further, many marketing constraints faced by farmers are interrelated. In terms of technology development, other marketing factors such as access to credit, adequate managerial and technical skills, and market information all influence technology choice decisions.

Oilseed Processing Technology Introduced by ATI

In the 1980s, ATI (now Enterprise Works Worldwide) started developing a manual ram press (also called the Bielenberg press after the engineer who developed the prototype) to produce edible oil in remote parts of Africa without electricity and with poor access to markets. The first presses were intended for soft-shelled sunflower seed in Tanzania, but the model has since been adapted for use with other oilseeds, including coconut and sesame. The first presses were arduous to use but later models improved on this and could be operated easily by one person. It was hoped that they would improve nutrition by improving access to energy-rich food, whilst also increasing farmer incomes and creating enterprise and employment in rural areas. Indeed, local entrepreneurs involved in small-scale oilseed processing can earn two to three times more gross income, compared with selling the seed to processing factories. In Tanzania, costs could be recovered in just one 3-month season. ATI worked up the promotion methodology over many years in many African countries. The ram press is quite simple and is supplied with a filtration device and tools for maintenance. Training, information and support on proper use and socio-economic and nutritional benefits of an oilseed processing operation are important elements of the extension programme adopted by ATI and its partners. In Tanzania, ATI sold the presses on credit for many years. Elsewhere, whilst varying in their systems and repayment rigour, credit has been an important factor affecting uptake. Depending on the model and manufacturer, the presses cost between US$ 150 and US$ 300. ATI has been innovative in promoting the commercialization of small-scale oil processing in Africa with its Regional Oils Programme. Its focus shifted from providing technical assistance to NGOs involved in small-scale oilseed processing to a private sector approach that emphasizes commercialization and mass manufacturing of ram presses. As a result, private enterprises have been developed, given assistance with market surveys and business plans, and are becoming responsible for the manufacture and wholesale of ram presses. ATI also stress the importance of the participatory processes used that involve economic actors in the community – owners and workers, press manufacturers, sales agents and oil-consuming families.

MARKETING LINKAGES

Marketing problems identified by producers are often attributed to the commercial sector and the capacity to access it: lack of buyers, unreasonably low farm-gate prices, inflexible requirements, and so on. Links between NGOs/ CBOs, the governmental sector and commercial agents are usually weak. Yet working together with the private sector is an important way for farmers to access relevant market information, technologies and new market opportunities. It is also a way for NGOs and CBOs to reach a wider audience with limited funds. As Kleih has stated, although NGOs are making a very positive contribution to rural development, they generally only reach around 1–2 per cent of farming households in any one country. Private sector agents are sometimes willing to collaborate with NGOs and farmer groups to share the costs of providing training and information, if they see it as an investment through which they can increase their own revenue.

In these circumstances, both the producers and traders can benefit. Developing and strengthening relationships between farmers and traders can reduce transaction costs, transport costs and risk on both sides. It was noted above that approaches that use and strengthen existing private sector production and marketing channels, rather than seeking to override them or invent new ones, have been particularly effective in addressing marketing constraints and in sustaining linkages beyond the life of the NGO marketing project. The Smallholder Agribusiness Development Promgramme (SADP) in Malawi, established by the American Co-operative Development Initiative and Volunteers in Co-operative Assistance (ACDI/VOCA), links farmers to existing private-sector markets, rather than establishing new market channels. This has contributed to its success in working with large numbers of farmers and achieving sustainability. CLUSA and

CARE have also focused on developing linkages between farmers and traders in Mozambique.

CREDIT PROGRAMMES

Credit merits particular attention in the context of agricultural marketing and processing. Farmers groups and NGOs often recognize a lack of credit as a critical constraint to the development of new initiatives and many seek to remedy this through credit interventions. The aspects discussed here are drawn from a recent synopsis by Gordon.

Rural Finance

Rural finance comprises credit, savings and insurance (or insurance substitutes) in rural areas, whether provided through formal or informal mechanisms. The word 'credit' tends to be associated with enterprise development, whereas rural finance also includes savings and insurance mechanisms used by the poor to protect and stabilize their families and livelihoods (not just their businesses). Rural finance comprises informal and formal sectors. Examples of formal sources of credit include: banks, projects and contract farmer schemes. Reference is often made to micro-credit. Micro underlines the small loan size normally associated with the borrowing requirements of poor rural populations, and micro-credit schemes use specially developed pro-poor lending methodologies. Rural populations, however, are much more dependent on informal sources of finance (including loans from family or friends, moneylenders, and rotating or accumulating savings and credit associations).

There is an enormous literature on rural finance and micro-credit, much of which relates to small-scale enterprise. (Many micro-credit schemes specifically exclude agricultural production activities, even in rural areas, because they are considered high risk.) The discussion here focuses on particular aspects of rural finance that are relevant to NGO and CBO agricultural marketing initiatives. First, typical features of micro-credit schemes, which tend to be run by NGOs and often work with community groups, are reviewed. Attention is then focused on two types of credit scheme with particular relevance to agricultural marketing interventions: inventory credit – a relatively new departure in smallholder agriculture, which is attracting increasing attention from NGOs and private banks; and outgrower schemes, which tend to be supported by the private sector but often work through producer groups. Some NGOs support other 'hybrid' rural finance initiatives that may affect agricultural marketing. These include stamp-based savings groups and rotating savings and credit associations (ROSCAs), where pooled savings may facilitate access to additional loans, and permit crop purchase/assembly/marketing activities or improved access to agricultural inputs.

Micro-credit and the Rural Poor: The Issues

Rural credit would not be the focus of so much development effort were it not for widespread market failure (*i.e.* failure to provide the services people want) in rural financial services in developing countries.

Reasons for market failure include:

- The lender does not know the default risk of each potential borrower and to collect this information is costly;
- It is costly to ensure that the potential borrowers take those actions which make loan repayment more likely;
- It is difficult and costly to enforce repayment;
- The cost of providing services to the rural poor is high because they are located in remote areas, want to borrow small amounts, and illiteracy, lack of experience of banks, and lack of collateral necessitate the development of tailored approaches.

What are the implications of this for agricultural activities? Firstly, all these types of market failure apply to agricultural lending in developing countries. Lack of information on the risk of default is particularly germane to agricultural enterprise. Farmers do not keep records, so it is difficult for them to produce the information that might convince a bank of their creditworthiness. Rural market transactions are largely informal, so it is difficult for the bank to collect independent information on prices. Farming is clearly a risky business because of weather, pests and market fluctuations and it is difficult for a bank to assess the degree of risk associated with particular activities. The rural poor do not have a track record, or referees who will vouch for their competence and reliability. Making sure that farmers keep to their business plan, using loans as intended, and carrying out tasks to schedule, is also costly – although this might make loan repayment more likely. Enforcing repayment is also difficult.

This requires monitors who know when crops are sold, or agreements with merchants to pay the farmer net of what she/he owes the bank, or effective penalties such as seizure of assets or prosecution. Farmers rarely have collateral acceptable to banks. They may not have clear title to the land they farm, or even if they do, rural land markets may not function well enough for land to be considered a 'bankable' asset. Poor farmers, moreover, rarely have other bankable assets. They might own a bicycle, and have a store half full of grain, but were a bank to seize such assets the cost of doing so would probably exceed their sale value.

The poorer the farmer, the fewer are his/her chances of borrowing from the formal sector. Women, particularly poor women, face even more problems in obtaining credit. Land title, where it exists, is usually held by men. Women often have little control over other factors of production, particularly for the 'bankable' cash-cropping activities. In some countries women may only borrow in the names of their husbands, if at all, and literacy rates for poor women are almost always lower than those for men. The irony is that numerous studies show that women tend to repay loans more reliably than men. Numerous projects, government schemes and NGOs engage in loan programmes targeted at the poor. Some of these work well, but many are unsustainable because of high and subsidized costs, and high rates of default. Moreover, many miss their target, with the benefits captured by the less poor. The poor depend overwhelmingly on the informal sector. Micro-finance is the response to market failure in 'conventional' banking services for the rural poor. It responds to the needs of low-income households.

Sound schemes tend to be characterized by:

- Small, short-term loans, and savings mechanisms;
- Simplified loan appraisal procedures;
- Innovative approaches to collateral;
- Rapid approval/disbursement of repeat loans after repayment;
- High transaction costs;
- High repayment rates;
- Savings and loan services provided at a location and time convenient to the poor.

Thus, micro-credit schemes are often associated with: group-lending (where peer pressure effectively substitutes for collateral, and other group members may take action to prevent one member defaulting, for instance, by providing labour to assure timely harvest); extension inputs arranged by the micro-finance institute (MFI); and mobile banking arrangements. Cashflow analysis may concentrate on overall ability to repay the loan rather than a particular investment project. In some respects, MFIs try to imitate the strengths of the informal sector (using local information to ensure repayment, for instance) and some MFIs are experimenting with ways to link their operations with some of the informal sector financial agents. Agricultural marketing loans are not common in MFI loan portfolios, but some micro-credit schemes do provide marketing finance or input loans. MFIs can include government and commercial banks, NGOs and savings and loan co-operatives. Some of the largest MFIs are Asian (for instance, Grameen Bank in Bangladesh has 2 million borrowers and savers, Bank Rakyat Indonesia/Unit Desa system has 2 million borrowers and 12 million savers, and the Bank for Agriculture and Agricultural Co-operatives (BAAC) in Thailand has 2.5 million clients). Coverage is less in Africa, but still growing. K-Rep in Kenya has 15000 clients and the Fe´de´ ration des Caisses d´ e´pargne et de Cre´ dit Agricole Mutuel (FECECAM) credit movement in Benin has 200 000 clients. Whilst these figures are impressive, they indicate a major difficulty: there are too few services to go round – and still fewer likely to become sustainable. Many schemes find it difficult to graduate from subsidized operations to full cost-recovery. Uncosted but critical inputs by NGO staff are commonplace.

Interest rates are often set below market rates (even at negative real rates, where high inflation prevails). Too little attention may be focused on repayment and follow-up; and farmer experience of subsidies, free inputs and loan amnesties may all contribute to low repayment rates.

Contract Farmer Schemes

Contract farmer or outgrower schemes focus on very specific needs. They usually operate in situations where a processor or trader faces a supply constraint and, therefore, wishes to promote production of crop x, and has access to sufficient resources (own or loaned) to do so. Inputs are usually provided in-kind (to reduce diversion to other activities), accompanied by extension, and the cost of these services is recouped out of the price paid to the farmer when the crop is harvested. The degree of supervision and type of inputs provided varies greatly. High value horticultural products produced for export may be closely supervised, for instance, whilst supervision of a cotton crop (which still has to undergo considerable processing before it reaches the consumer) may be minimal. Viable schemes must include mechanisms to minimize on-farm consumption or side-selling (farmers selling their crop to an alternative buyer and, therefore, avoiding loan repayment), such as:

- Sharing of information and co-operation amongst crop buyers;
- Effective penalties against farmers who default (exclusion from future schemes, seizure of assets, prosecution);
- Peer pressure achieved through group lending.

Some of the longer-established cotton schemes in West Africa are wide ranging in scope. Considerable long-term effort has gone into capacity building of farmers' groups. This enables farmers to take on a greater share of tasks in input supply and crop assembly, which both reduces cotton company costs and increases cash revenues to farmers. Inputs and extension are available for other crops in the farming system (not just for cotton) effectively reinforcing and expanding the benefits accruing from cotton production.

These types of schemes are common for smallholder annuals grown for export such as tobacco, cotton and horticultural production, where the on-farm investment costs are relatively low, and the pay-back period short. (Thus, greenhouse production of flowers for export in Zimbabwe is not a smallholder activity.) In Africa, the cotton schemes are undoubtedly the largest. About 300000 Ugandan farmers benefit from an input scheme organized by the cotton ginners. In Zimbabwe, around 60000 communal farmers take input loans from cotton companies (and many more participate in other schemes intended to increase cotton output). In Mali, around 100000 rural households participate in the cotton input schemes. The schemes in Mali and Zimbabwe both depend on a network of strong farmers groups, and the Ugandan scheme is likely to develop in that direction as it matures, seeking less costly ways to reach its target group.

Crop Market Characteristics

If there is limited on-farm consumption and local marketing, it may be possible to collect loan repayments at the point of sale. However, unless there is a crop purchase monopoly, or agreement to share information between different buyers, the farmer may be able to avoid repayment by 'side-selling' (selling to another buyer).

Input qualities

Inputs are often provided in-kind to reduce 'diversion' of the input away from the targeted crop. Diversion is lessened if there is a limited alternative use or market for the inputs, or if (unusually) returns to use of the input are greatest for the crop in question.

Commercial/Credit Context

Prospects for viable operation of the scheme are greater if farmers treat farming as a business and are integrated into markets, and if there are supportive legal/political/ contract enforcement institutions. A recent history of loan amnesties, default without penalty, and subsidized inputs may undermine the operation of viable schemes.

Modus operandi of scheme – best practice:

1. Group schemes for peer pressure.
2. Group or individual schemes backed up by monitoring/good information, support staff, and ability to act.
3. Incentives for repayment and penalties for non-repayment.
4. Appropriate incentives for field monitors/co-ordinators.
5. Training provided to farmers extension and business management.
6. Developing relationship/trust/loyalty through field presence/contact.
7. Accessibility of scheme – minimize red tape and transaction costs; organize so that the location and timing of contact is convenient to farmers.
8. Effective and timely monitoring of input use and crop marketing.

Inventory Credit

Another mechanism for financing agricultural trade exists in inventory credit or warehouse receipt systems. This involves a tripartite agreement between a bank, a borrower (usually a trader), and a warehouse operator. Essentially, a trader is able to use an existing stock of, say, grain as collateral for a bank loan providing certain conditions are met. The grain must meet certain verifiable specifications and be stored in a warehouse operated by a third party.

The bank will usually lend up to a certain percentage (perhaps 80 per cent) of the value of the grain at that time (at harvest time when prices are low). The trader can then use the loan to acquire additional stocks of grain, and is usually required to repay the loan during the 'lean' season before grain prices start to fall once more with the next season's harvest.

The trader must settle all the warehousing costs before she/he removes the grain, and is generally required to repay the loan at this stage also. In the event that she/he is unable to repay the loan, the bank can seize the grain and sell it.

There is a lot of interest in inventory credit at the present time because of the scope it offers to provide traders with capital to fuel agricultural marketing. However, its use is limited by a number of factors: it can only be used for crops that are relatively non-perishable, with reasonably predictable pricing scenarios; by definition it is targeted to larger, more sophisticated traders, able to purchase initial stocks and fairly confident in their negotiations with banks and warehouse operators; and it is critically dependent on the necessary institutional infrastructure (bank willingness to lend against inventories, warehouse systems which can operate to the standards and within the necessary legal framework, and supportive and enforceable legal institutions). NGO involvement in these schemes can take a number of forms. It may include a role as an MFI providing the credit, or it may include working with farmers' groups who effectively become the 'trader' in the tripartite agreement, buying crops from other farmers (or their membership), for storage and sale later in the season when prices are higher.

AGRICULTURAL MARKETING AND AGRI-BUSINESS DEPARTMENT

STATE SCHEME

Own building for Agmark Laboratories, Strengthening of Agmark Laboratories and Provision of Computers (₹ 3.07 crores): There are 30 State Agmark Grading Laboratories, one Principal Laboratory and 15 Agricultural marketing centres functioning in the State. It is proposed to construct building for Agmark Labs as well as modernize the equipments during the Tenth Five Year Plan for which a sum of ₹ 3.07 crores is provided.

Schemes Proposed to be Implemented Availing Financial Assistance from the other Financial Institutions

Provision of Infrastructure Facilities (₹ 100 crores)

i. *Creation of infrastructure facilities:* There is a need to provide infrastructure facilities like transaction shed, drying yard, farmers rest sheds, sanitation facilities, drinking water supply electronic weighing scales, cleaning and washing facility, moisture meters, scientific instruments for grading, rural godowns etc., for which an amount of ₹ 100 crores is provided.

ii. *Revamping Tamil Nadu Agricultural Marketing Board:* There is an imperative need to revitalize the Board activities by making this body more responsive and independent decision- making body. It should be given administrative and financial powers.
iii. *Revitalising the market committees:* At present the Act provides for nomination of members to the Market Committees. To make it more accountable and democratic, it is necessary that members be elected so that there is a sense of belonging. Secondly, there is lack of professionalism in the staff of the Market Committees. Their mind set is more of regulation than market oriented approach. Training to the existing staff and inclusion of professionals is a must.
iv. *Provision of infrastructure facilities in Post Harvest Centres:* At present the post harvest centres in the Regulated markets conduct training programmes to the farmers and exhibitions on various post harvest practices. The regulated markets should expand their activities through provision of whole storage CA, MA, Retardation and ripening facilities for the agricultural produce.
v. *Agricultural Marketing Extension:* Strong network of marketing extension is very much necessary at block level to effectively advise farmers on various aspects of marketing, advise on product planning, marketing information, and securing market for farmers, and advise on improved market practices and advise on post harvest management practices.

To strengthen the regulated market it is proposed to provide infrastructure facilities like transaction shed, drying yard, farmers resting shed, sanitation facilities, drinking water supply, electronic weighing scales, cleaning and washing facility, moisture meters, scientific instruments for grading, rural godowns etc. at a cost of ₹100 crores with financial assistance from Marketing Committee, NABARD, GOI.

Provision of infrastructure facilities in Post Harvest Centres-Cold Storage (₹ 200 crores- National Horticultural Board/Private Sector)

It is estimated that around 30 per cent of the horticulture produce is wasted due to inadequate cold chain facility and appropriate technology for the preservation of horticultural produce. At present the combined cold storage capacity of 133 units in Tamil Nadu is around 1 LMT. The existing capacity is not sufficient to store horticultural, dairy, and marine products. In order to meet the demand it is proposed to establish an additional 1 LMT capacity of cold storage facility, at an estimated cost of ₹200 crore. Taking into consideration the initial capital investment, high recurring cost and low capacity utilisation and project failure, it has been proposed to attract investment by private sector through measures like financial incentives in addition to NHB's subsidy and power tariff concession.

Agricultural Marketing Extension/ Packaging Training (₹ 42.95 crores)- Tamil Nadu State Agricultural Marketing Board (TNSAMB)

Agricultural farmers require advice on various aspects of marketing like selection of the crops to be grown with marketability in mind, current price, market arrival and forecasting of market trends, and on post harvest management practices.

Agro-processing Food Park (₹ 10 crores)- Private Sector

It is estimated that about 10-20 per cent of the food grain production and 30-35 per cent of fruits and vegetables are wasted at various stages from picking to consumption. This indicates the need to establish food-processing industries to preserve and minimize the wastage. To encourage private sector to establish agro-processing industries, the Government of India, Ministry of Food Processing Industries has formulated two models of Food Park for Food Processing Industries apart from other incentives. It is proposed to set up a Food Park at a cost of ₹ 10 crores either in private sector or by Market Committees.

Setting up of Terminal Markets and Collection Centres (₹ 160 crores)

Private sector or growers association in partnership with private sector can organise Terminal Markets for specified products with backward integration with collection centres (value addition centres). Fifteen to twenty collection centres established nearer to the production area, can feed one Terminal Market. Two such a Terminal Markets will be established one each at Chennai and at Coimbatore with NDDB assistance.

Mega Markets (₹ 10 crores)

A mega market centre can cater to the needs of wholesale dealers, exporters and food processing industries. The availability of horticultural products in large scale in one place would promote exports and provide single sourcing to agri processing industries. It is proposed to establish a mega market for vegetables at Oddanchatram in Dindigul District at a cost of ₹2.59 crore. Depending on the success of this project, mega markets could be developed at Mettupalayam in Coimbatore District and Athur in Salam District either through Market Committees or private sector.

Agro Processing Industries (₹ 10 crores)

Food Processing Industries provide the crucial farm - industry linkages, which help to add value to the produce, generate employment opportunities and increase the net income to the farmers. The Market Committees will establish these through private sector in 10 places with an incentive of 20 per cent of the capital cost. The project cost is estimated to be ₹10 crores. Packaging fresh fruits, vegetables and other farm produce is an important process as it reduces post harvest losses, increases the income of the farmers and ensures clean and hygienic farm produce to the consumers. It is proposed to create awareness and impart training on farm produce packaging to the technical officers of Agriculture, Horticulture, Agricultural Marketing and Agri Business departments during the Tenth Plan period. It is proposed to train 600 officials and 400 farmers at a total estimated cost of ₹25 lakhs. It is also proposed to conduct studies for developing low cost packaging techniques and material for some specific commodities at a cost of ₹0.35 crore.

Agri Export Zone (₹ 150 crores) –GOI

The most critical factor to meet the challenges of export will be enhancing exporting capability of our State in a highly competitive environment. To promote export of agri products, Government of India, Ministry of Commerce has announced in 2001 - Exim Policy, a scheme of establishing Agri Export Zone (AEZ). In AEZ, institutional and physical infrastructure would be created as per the needs of the specific commodity. Steps have been taken to establish AEZ for cut flowers, mangoes, bananas, medicinal plants and vegetables. To promote agricultural exports, it is proposed to educate and train the growers of identified crops in producing, grading and packing for international market, and in establishment of analytical laboratories, setting up of Export promotion cell in Agri Business department to disseminate information and harmonization of standards of Indian products with international standards (Codex).

MARKETING EXTENSION

Network is proposed to be formed integrating with the extension network already available with Agriculture Department. Officers of Agriculture, Horticulture and Agricultural Marketing departments will be given training on various aspects of Agricultural Production and Marketing for the purpose of carrying out extension works effectively and efficiently.

A sum of ₹65 lakhs is proposed during the Tenth Five Year Plan. Further, the Government will also facilitate private sector to carryout extension work for the quick reach of farm information to the farmers. To improve market practices and also sale of specific agricultural products, two products' specific market Complexes *viz*- Turmeric Market Complex at Erode at a total cost of ₹ 32.30 crores, and Jaggery Market Complex at Trichy at a total cost of ₹ 10 crores will be established. These market complexes will be established incurring expenditure initially from Market Committees and financial institutions like NABARD and then recovered from the traders in instalments.

ALTERNATIVE MARKETING FORMS

Role of Government in managing markets is on decline worldwide. It is not easy to bring major changes in the traditional marketing system. The only way to modernize marketing is to promote alternative marketing system and that may operate parallel to and in addition to present marketing system. The purpose of the proposed alternative marketing is to promote modern trade practices, which in turn will pave way for transperancy and effiecncy in market.

Even though, the various forms of alternate marketing like:

- Direct marketing,
- Marketing through farmers interest group,
- Setting up of terminal markets,
- Forward and future market,
- e-commerce,
- Setting up of mega markets,
- Negotiable warehouse receipt system etc. have been suggested by Expert Committee on Agricultural Marketing headed by Shankarlal Guru, three important marketing methods could be considered in the State *viz.*, Terminal Market, Mega Market and Direct Market.

AGRO PROCESSING INDUSTRIES

Food Processing Industries provide the crucial farm - industry linkages which helps to accelerate overall agricultural development, adding value to the produce, generating employment opportunities and increasing the net income to the farmers. In India, only 2 per cent of the total horticultural produce is processed. In countries like Brazil, it is in the range of 70 per cent. Considering the rising demand for good quality products, there is an urgent need to enhance capacities for value added and processed products. At present, value addition is estimated at 7 per cent of the total production within next 5 years.

There is a need to increase value addition to 20 per cent and processing at 7 per cent. To increase our country's share in the world trade of agri products which stands at less than 1 per cent at present, the most critical factor in the highly competitive market environment is quality processed products. In Tamil Nadu, food processing in the form of drying, vegetable oil, grain processing, sugar breweries are in existence for a quite long period. Lack of adequate infrastructure facilities like storage, processing, marketing besides technical know-how have been the major constraints affecting the growth of the industry. Tamil Nadu with a coastline of 922 km and surface boundary of 1200 km with tropical and sub-tropical climate is ideally suited for the production of a host of agricultural, horticultural, acquacultural and animal husbandry produces. Thus there is vast scope for setting up of Food Processing Industries in Tamil Nadu.

The growth of food processing industry, which is included in the priority lending sector, will bring immense benefit to the economy. Economic liberalization and raising consumer prosperity is opening up new vistas in food processing sector. Tamil Nadu produces 103 lakh metric tonnes of food grains, 22 lakh metric tonnes of oil seeds, 37 lakh metric tonnes of sugarcane, 53.89 lakh metric tonnes of fruits, 54.65 lakh metric tonnes of vegetables, 2.95 lakh metric tonnes of spices and 6.94 lakh metric tonnes of plantation crops. 10 to 20 per cent of food grain production is lost every year both at pre and post harvest stages. Similarly 30 to 35 per cent of fruits and vegetables produced are also wasted at various stages from picking to final consumption. This only indicates the immense potential and scope that exists for setting up of food processing industries so that the available raw materials may be put into the most judicious use. This also indicates the scope that exists for adoption of post harvest modern techniques ensuring preservation and minimizing wastages of agricultural produce.

IMPROVEMENTS IN THE PRODUCTION AND MARKETING

Marketing projects were limited to products which were subject to common price and market regimes but, as the range of commodities within the price policy expanded, this was no obstacle. Article 12 provided an interpretation of the four categories listed in Article 11 and it is interesting, even at such an early stage in the history of the CAP, that 'the adaptation and guidance of production' meant 'the quantitative adaptation of production to outlets' and 'improvements in the quality of the products'. During the 1960s—and indeed for a large part of the 1970s—Reg. 17/64 provided the main source of funds for structural reform. The range of activities covered by the Regulation was very wide, encompassing production and marketing projects, and those which were a mixture of the two. Over the lifetime of the Regulation, projects to improve production structures received 53 per cent of the funding, marketing structures 41 per cent, and mixed projects 6 per cent.

The Community contribution to the investment was initially very low at 25 per cent for all projects but, in 1973, this was raised to 45 per cent of eligible expenditure in projects related to production, with the beneficiary only having to provide a minimum of 20 per cent of the cost (originally a minimum of 30 per cent). Thus, right from the beginning, the Community element in structural legislation was low, unlike the price and market policy where, by the end of the 1960s, FEOGA-Guarantee was providing 100 per cent of the support. However, it should be noted that, for structural measures generally, the tendency over time has been for the Community contribution to rise, particularly in the poorer regions or in the poorer Member States.

During its lifetime, Reg. 17/64 was extremely popular and far more projects were submitted for funding each year than were ultimately aided. The sectoral distribution of aid, 1964-1979. Three sectors —milk, wine, and meat—accounted for one-third of the aid; land improvement and hydraulic works accounted for about another third; and all other purposes the remainder.

The result was that the better-off farmers and the more prosperous farming regions were benefiting disproportionately from all aspects of the

CAP. An interdepartmental group within the Commission reported *inter alia* on this issue, and its findings in relation to the regional impact of Dir. 72/159 have already been discussed. As to the regional impact of Reg. 17/64, if it had been operating as originally intended, each Community Programme would have had specified zones of principal action within which investment aid would have been concentrated. The Commission had estimated that, as a general rule, these zones would have represented about one-third of the designated area or other appropriate measure, such as volume of production.

In reality, in Reg. 17/64, there was no provision for regional differentiation of aid, rather the funds available were allocated to the Member States according to a prearranged system of quotas. Thus, prior to Enlargement, Italy received annually an allocation of just under 34 per cent of available credits, Germany 28 per cent, France 22 per cent, and so forth. These percentages were adjusted downwards to accommodate the three new Member States after 1972.

The interdepartmental group was interested to examine how these national quotas had operated in practice and to compare them with the percentage distribution among the Member States of: (a) the agricultural labour force; (b) final agricultural production; (c) the simple average of (a) and (b), which in effect gave production weighted by the inverse of agricultural labour productivity; and (d) agricultural area. In no case did the quota exactly match any of these measures, although it came reasonably close to (c). For some countries the divergence was considerable—which was hardly surprising, as the quotas had been devised more on a basis of horse-trading than on any economic or social indicator.

Within some of the Member States, the group discovered that the quotas were apparently being divided by region, according to the size of the agricultural area. This meant that 'some of the relatively extensive and well-structured agricultural regions of the community (such as Schleswig-Holstein, Niedersachsen, Scotland, Emilia-Romagna, Toscana and Lazio) receive much higher expenditure than agricultural employment or structural criteria would suggest'. The group also commented on the considerable difference between expenditure commitments for Italy and actual payments: in 1964-72 it received about one-third of total commitments but only half that amount had been paid by 1974. Thus, even at this early stage, administrative difficulties were preventing Italy from gaining access to structural funds—a pattern that was to become familiar as the years went by.

It is of some interest to examine whether or not the perceived lack of balance between the need for structural improvement in poorer regions and the distribution of total payments from Reg. 17/64 persisted over the lifetime of the scheme. In 1989, the Commission published a cumulative regional breakdown of the aid granted. While Italy received the largest payments, and while the South and the islands received a considerable share of the total, Emilia-Romagna remained a main beneficiary, and Lazio received more than Sicily.

Germany, which received the second highest payments, still had Niedersachsen as its second highest recipient, although Schleswig-Holstein did not maintain its earlier position. In the UK, Scotland remained by far the largest recipient. In Belgium, the south is poorer than the north but received less aid, while in Ireland two regions—South-East and South-West—dominated, although the West and North-West were much poorer.

Member State of the total number of projects financed. Most were already completed by the end of 1992. About 11.5 per cent had been abandoned, for whatever reason, by far the greatest number of these were in Italy. Despite its late arrival, the UK made very good use of Reg. 17/64. Member State of the aid granted, cumulatively to the end of 1992. This should be compared with the payments made. While the sources are not the same, which may have resulted in some discrepancies, in broad terms one can conclude that the account is nearly closed on this scheme. In Italy there is a considerable difference between aid granted and payments received, which presumably is a reflection of the large number of projects which were abandoned.

So the life of this extraordinary piece of legislation draws to a close. For so many years, it provided a link with the earliest days of the CAP and, in the distortions which so quickly developed in its operation, was a reminder of how soon in the life of the policy the politics of the CAP usurped the ideals of the original designers. However, some aspects of Reg. 17/64 have remained. In 1976, the interdepartmental group referred to earlier recommended that, when Reg. 17/64 ceased, its infrastructural aspects should be restricted geographically in any replacement legislation.

In a sense, this did occur with the measures which came into force in the late 1970s and early 1980s. However, it was not until the reorganization of the Structural Funds in the late 1980s that a more systematic approach to regional concentration of effort was introduced into the CAP. The aid to processing and marketing was carried forward with the enactment of Reg. 355/77.

THE ESTABLISHMENT OF PRODUCER GROUPS

One of the most obvious features of agricultural production in Western Europe (and in many other parts of the world) is its organizational structure, dominated by small and medium-size independent farms. Production in such a framework comes very close to the economist's definition of a perfect market. Farmers are price-takers: they are very rarely in a position to dominate the market, even if they are organized into a commodity or special interest group. In general, they produce a homogeneous product and therefore cannot benefit from higher prices achieved through product differentiation.

They are in a weak bargaining position, whether selling a product for direct consumption or for processing, and this is particularly true when the product is perishable (as with milk) or perishable and seasonal (as with fruit and vegetables).

The poor structural organization of agriculture was even more pronounced in the 1960s in the new Community than it is in the 1990s. Farming was characterized by a very large number of exceedingly small producers, each with an extensive range of products. This situation was not conducive to orderly marketing, particularly at a time when consumer demand was changing rapidly due to the marked increase in living standards.

What was required was regular supply, known and constant quality, and stable prices. It was essential that producers respond and that they took greater responsibility for the marketing of their own products. Vague references were made by the Commission as far back as the 1959 and 1960 proposals on the CAP to its intention to encourage initiatives by farmers' organizations, which would lead to better information on markets, to improve their stability, and their response to consumer demand.

The Commission reasoned that, if producers of the same commodity came together in an alliance, it would enable them collectively to improve the quality of their product by enforcing grading standards, and the continuity of supply, by having centralized storage and processing facilities. The group would market the output of its members, thereby improving their bargaining power and, as a result, obtaining a higher unit price. This in turn would help to further the aims of Art. 39.1 of the Rome Treaty, in particular as regards incomes, market stability, and supply availability.

These thoughts were carried a stage further in Reg. 26/62 on the application of certain rules of competition to production of and trade in agricultural products. Under this Regulation, farmers and their organizations were, in general, excluded from the ordinary Community rules on competition, in so far as their activities were concerned with the production and sale of agricultural products, and the use of common facilities for the storage and processing of such products.

Although, theoretically, the case for collective action is strong, in practice a recurring difficulty with any sector dominated by small-scale independent producers is to persuade them to join a group which will have power to impose standards, control prices, and even production, and once they have joined to prevent them from breaking their contractual obligations to the group, thereby undermining its collective bargaining power.

One way in which the advantages of membership of a producer group can be enhanced is to give to such groups certain formal market intervention responsibilities. Prior to the establishment of the European Community, the Netherlands provided good examples of how such a system functioned, especially in the fruit and vegetable sector. The characteristics of the Dutch system were outlined in the 1960 proposals on the CAP, and certain features of that system found an echo in some of the commodity regimes. The first example came in Reg. 159/66 *concerning additional arrangements for the common organization of the market in the fruit and vegetable sector*. This Regulation completed the arrangements for fruit and vegetables which had been hanging fire since 1962. A novel feature of this regime is the important role played by producer groups which act as the agency for internal market support.

For many commodities, provision is made in the relevant market regime for the withdrawal of produce from an oversupplied market, its storage, and disposal. These tasks are normally undertaken by an official agency or agencies established in each Member State. However, for a small number of products, market management is in the hands of the producers themselves, through producer groups which receive official recognition. Given the characteristics of fruit and vegetable production—a highly perishable product, many small-scale producers, local markets, seasonal production, volatility in supply and price-flexibility and speed of response are required if any form of market management is to be effective.

Apart from market support for certain commodities, more generally producer groups provide their members with facilities for handling and marketing the output of the relevant commodity. In their turn, the members are expected to market this total supply through the group (unless the group allows otherwise), and they must abide by the rules adopted by the group for the improvement of quality and the adaptation of supply to demand. The involvement of FEOGA-Guidance lies in the financial encouragement it provides for the establishment of producer groups. The first example of such aid appeared in Reg. 159/66 under which Member States were given the possibility of helping producer groups to get started, by providing them with a degressive aid during the three years following their establishment. This annual aid was not allowed to exceed 3, 2, and 1 per cent respectively of the value of the production marketed by the group. The Guidance Section reimbursed 50 per cent of the Member States' eligible expenditure under this scheme. Tentative references were made in Reg. 122/67 *on the common organization of the market in eggs* and in Reg. 123/67 *on the common organization of the market in poultrymeat* to the need to encourage efforts to improve the organization of production, processing, marketing, and quality through trade groups. However, in these commodity regimes, no financial support was offered to help launch such groups.

In the light of all this piecemeal and unco-ordinated interest in producer organizations in various commodities, it is not surprising that the Commission eventually proposed the establishment of a Community framework within which Member States could encourage the formation of producer groups in any commodity. The Commission was also prompted to do so because certain Member States were already promoting these groups and there was a danger that problems would arise at Community level through the use of different criteria and objectives in the various countries concerned.

Thus, in 1967, the Commission proposed a Regulation on producer groups and their unions in COM(67)68 final, which set out common rules on the type of aid which could be given by the Member States to encourage the formation and development of these groups, and on the conditions under which they could be officially recognized. Despite the Commission's evident interest in the creation of producer groups and despite the existing Community launching aids in the fruit and vegetable sector, the 1967 proposal did not make provision for Community aid. This was somewhat surprising and indeed was criticized by the European Parliament in its Resolution on the draft.

Although Mansholt always regarded producer groups as an extension of market organization rather than as a part of structural policy, his 1968 Plan did contain a discussion of the need to improve marketing, and one of the measures suggested to aid that process was the immediate adoption by the Council of the proposed Regulation on producer groups. The Council clearly did not heed this exhortation because, in its 1970 structural reform proposals, the Commission made a pointed reference to the thorough discussions which had taken place in the European Parliament and in the ESC on the 1967 producer group draft and stated that the Council had not even begun a discussion on it. The 1970 reform proposals included a considerably revised draft Regulation on producer groups—the most important change being the provision of Community aid to help launch the groups. In the explanatory note to the proposals, the Commission made it clear that it regarded the improvement of market structure as primarily the responsibility of the Member States. For that reason, it proposed that the Community should refund only 30 per cent of eligible national expenditure. The logic of this position is unclear, as it will be recalled that the Community was already refunding 50 per cent of eligible expenditure on producer groups in the fruit and vegetable sector.

In 1971, when the revised structural reform package was presented to the Council, the draft Regulation on producer groups appeared again—now in a third version. The changes made were generally of a technical nature but, importantly, the Community contribution to eligible expenditure by the Member States had been reduced to 25 per cent. It was all to no avail, as the draft Regulation was not passed by the Council.

In the *Resolution* on the new orientation of the Common Agricultural Policy in 1971, the Council finally agreed in principle to the introduction of a regime to encourage the formation of producer groups but, once again, it appeared that Community aid would not be offered and that the funding would be by the Member States. Despite this, however, the Council remained unable to agree to the proposal, although discussions dragged on into 1972. The only progress which was made was under Reg. 1696/71 *on the common organization of the market in hops.*

The hops producer groups are not involved in price support operations but are intended in a more general way to assist in the centralization of supply and in the adaptation of the product to market requirements. Member States were free to grant a launching aid and FEOGA-Guidance reimbursement was limited to 25 per cent of eligible expenditure. An extra feature of this regime was the option given to Member States to aid producer groups on a temporary basis to reorganize hop gardens and to switch to more suitable varieties. The form of aid was a payment per hectare replanted or reorganized, and FEOGA-Guidance refunded 50 per cent of the eligible expenditure of participating Member States.

Discussions resumed in 1976 on the introduction of a general producer group regime and, in the following year, the Commission put forward yet another revised proposal. This was so radically different from previous drafts as to be, in effect, a completely new proposal. In the explanatory memorandum, the Commission tried to explain why such a harmless piece of legislation was causing such difficulties.

Basically the problem was that certain Member States with well-developed production and marketing structures could not see the necessity for this legislation. If anything, the differences between the Member States had grown since the original 1967 proposal. The Commission, therefore, proposed that instead of a Regulation of universal application, there should be a measure which would apply only in those areas where the supply of agricultural produce was structurally very inadequate. Primarily this meant Italy and indeed the proposal referred solely to that country, with the provision that the Regulation could be extended to other Member States.

The Council had agreed in February 1977 that it would adopt the proposal by the end of June. It was in fact a full year later that Reg. 1360/78 *on producer groups and associations thereof* was enacted. It was almost totally unrecognizable from the draft and covered not just Italy but also Belgium, Southern France, and DOM (the French overseas departments). The commodity coverage differed by country, and in France by region. It was intended to be of assistance to farmers in regions and sectors where there were severe structural deficiencies, due to the small size of farms and poor organization. For a measure which was to last initially for five years only, at a total FEOGA cost of 24m. ua, an incredible amount of bureaucratic effort was involved. When Greece, Portugal, and Spain joined the Community, Reg. 1360/78 was extended to them, and to Ireland and the remainder of France in 1988, all for specified lists of commodities.

The only other commodity for which producer group launching aids are available is cotton under Reg. 389/82. Under the terms of Accession, when Greece joined the European Community, market support was extended to cotton and producer groups were seen as an important mechanism for the rationalization of production and the centralization of handling and marketing. For this reason, the Regulation covers not only launching aids but also investment aid to assist with structural improvements in harvesting, ginning, storage, and packaging. Producer groups in Italy and Spain are also eligible for aid under this regime. Having sketched the rather confused history of the development of legislation on producer groups, the remainder of this Section reports on the statistical evidence of the success or failure of the various structural measures to persuade farmers to come together to improve the organization of particular commodities. Data on structural aids are available for producer groups in fruit and vegetables, hops, cotton, and miscellaneous products under Reg. 1360/78.

In 1972, Reg. 159/66 was replaced by Reg. 1035/72, which consolidated all the existing Community legislation on fruit and vegetables. Arrangements for launching aids for producer groups were retained unchanged. The pattern which is so familiar from other structural aids is present here also: aid is concentrated in a very small group of countries and it is not necessarily related to the size of the sector. For instance, the very small number of producer groups set up under this legislation in Belgium and the Netherlands is indicative of the strength of the previously long-established marketing structures based largely on the cooperative movement. Therefore, there was little incentive or need to seek launching aids. In contrast, the relatively large number of launching aids in the UK is an indication of the previous absence of joint ventures by farmers rather than an indication of the size of the horticultural sector. The size of producer groups varies considerably from one Member State to another, as evidenced by the absence of any relationship between the number of groups and the level of investment.

Hops provide a real contrast to fruit and vegetables. This is a very small sector, highly localized in traditional areas of production where, of course, it can be quite a significant crop. Much of the output is grown on contract. The schemes of aid came to an end some years ago and 1985 was the last year for which data were available. The situation with regard to cotton is quite different. Greece is the largest producer and so far (at the time of writing) was the only Member State to have made use of the launching and investment aids, although the Commission anticipated that Spain, which is the only other significant producer, might well begin to show an interest in this measure.

In 1991, the Commission reported to the Council on the operation of Reg. 1360/78. By the end of 1990, the Member States had granted recognition to 560 groups. The discrepancy is explained by 'the slowness of the Member States' administrations in drawing up implementing rules and requesting refunds'. Despite the fact that Belgium was included in the scheme right from the start, no groups in the relevant production sectors (cereals, cattle, pigs, and Lucerne) had applied for recognition, even though the requirements had been made easier.

Indeed, the Commission stated that the Belgian authorities had reported 'that those concerned are uninterested in the very idea of producer groups and that, in the present circumstances; this is unlikely to change in the future'. For that reason, the Commission proposed that, when the life of the Regulation was prolonged, Belgium should be excluded. This was rejected!

IMPROVEMENTS IN PROCESSING AND MARKETING

The fact that the processing and marketing support activities of Reg. 17/64 were continued in Reg. 355/77 on common measures to improve the conditions under which agricultural products are processed and marketed. This Regulation was a long time in coming. The life of Reg. 17/64 was limited following the introduction of the financial Regulation 729/70, which laid down the basis for future assistance of a structural nature. If for no other reason than that of legal necessity, consideration had to be given to the replacement of Reg. 17/64 by some other measure in the sphere of marketing and processing.

More than that, however, there has always been a tendency to concentrate aid to agriculture in that sector of the industry which lies on the farm and to neglect the need to improve the handling of the products once they leave the farm. For many years, Reg. 17/64 was the only Community measure of any importance providing such post-farm aid, despite the fact that it was recognized that the agricultural sector was weak in the handling of produce.

In setting the new guidelines for the CAP in 1971, the Council had invited the Commission to continue to study the problems of marketing and processing, and to submit a proposal on their alleviation. In 1973, in the *Improvement* memorandum, the Commission announced its intention to make such a proposal but it was not until mid-1975 that it finally managed to submit this proposal to the Council in COM(75)431. The Commission reasoned that a more efficient processing and marketing sector would be in a position to pay better prices to producers, to diversify output, thereby stimulating demand, to concentrate more on exports, and (rather obscurely) to 'better handle products from remote areas of the Common Market'.

The proposal was for the establishment of multi-annual programmes, each of which would cover one or more agricultural products. Within these programmes the Community would provide aid for suitable projects intended to promote the rationalization and expansion of processing and marketing. Assistance was to take the form of a capital subvention of not more than 25 per cent of the value of the investment, with the beneficiary providing at least 50 per cent and the Member State participating in the funding as well. The estimated cost of the scheme was 400m. ua over a five-year period.

While largely in agreement with the intentions of the Commission, the ESC had a number of reservations.

For one, the Committee was of the opinion that the proposed Regulation was 'of direct concern to consumers, farm workers, and workers in the food industry and the distributive trade' but that insufficient regard was being paid to the interests of these parties.

The ESC also believed that the effectiveness of the measure could be improved by the formation of producer groups and urged the Council to adopt the long out standing draft Regulation on such groups. This point was echoed by the European Parliament.

Another point on which the two bodies agreed was their scepticism concerning the Community's ability to control the new measure in a manner which would avoid exacerbating the surplus problem. Both Parliament and

ESC considered the proposed scheme as financially inadequate. The Parliament stated that 'the Commission's proposal represents a limited step which will result in a decrease in the total real amount of Community aid to be granted for the improvement of marketing and mixed production/ marketing structures, and which will make no substantial contribution to reducing agricultural surpluses and limiting the need for intervention'.

The ESC commented that '80 million units of account per year is not enough to solve all the problems of improvement of agricultural product marketing and processing facilities, especially as costs will continue to be pushed up by inflation'. It was not until 1977 that a modified version of this proposal was passed by the Council and it is hard to believe that either the Parliament or the ESC would have regarded the changes as beneficial.

The programmes 'to develop or rationalize the treatment, processing or marketing of one or more agricultural products' had to be drawn up by each Member State and agreed by the Commission before applications for the funding of individual projects could be considered. These projects could be for a wide range of activities: the development of new outlets for agricultural products; lightening the burden on intervention by improvements to the structure of a market; assistance to regions having difficulty adjusting to some aspect of the CAP; improvement of marketing channels or rationalization of the processing of an agricultural product; improvement of quality or presentation... While the contribution from FEOGA-Guidance to eligible expenditure was fixed at 25 per cent, in later years this was modified in various countries and regions, *e.g.* in the Mediterranean, and the West of Ireland. Strange little schemes were added on from time to time as, for instance, a special measure to rationalize and improve slaughterhouses in Belgium.

The support arrangements for processing and marketing were overhauled in 1990, so as to bring them into line with the new arrangements for the Structural Funds.

However, in some Member States there have been shifts, with new regions coming to the fore, such as the North East in Ireland. There still was evidence that the poorest regions were not achieving the enhanced level of investment necessary for them to narrow the income gap. There were great differences by country, with one or two sectors in each dominating: meat in Belgium; fish in Denmark; cereals and wine in Germany; fruit and vegetables, and tobacco in Greece; fruit and vegetables, and meat in Spain; wine, and fruit and vegetables in France; meat in Ireland; fruit and vegetables, and wine in Italy; fruit and vegetables in Portugal; and fish and meat in the UK. Taking the Community as a whole, fruit and vegetables, meat, and wine projects dominated in terms of number, and together they also accounted for over 57 per cent of the funds. As with Reg. 17/64, the outstanding feature of the table is the relatively large number of projects abandoned in Italy—over 38 per cent of the total. The figure for the UK—although much lower—is also rather high. The second part of the table gives the cumulative figures for funds committed and actually paid. The rate of reimbursement is an indication of the efficiency of the programme in the Member States, both in terms of the take-up of the investment opportunities and of the national bureaucracies. Five Member States had achieved a payment/commitment level in excess of 70 per cent; France and the UK achieved levels in excess of 60 per cent; Ireland and Italy in excess of 50 per cent. Greece, Portugal, and Spain were well below these levels. While the last two countries had only recently joined the Community and therefore could be expected to be somewhat slow the same cannot be said for Greece.

In terms of total funds, aid to marketing and processing is one of the biggest sources of structural investment in agriculture—just as Reg. 17/64 had been before it. Together with funds spent on improving the efficiency of farms, and LFAs, these three are the most significant programmes. While aid to LFAs has a large social element in it, in that it is intended to compensate for inherent locational handicap, the other major aids are intended to make lasting improvements to the economic framework within which farming takes place and agricultural produce reaches the market.

Despite their popularity, one must question to what extent they have been successful in raising levels of efficiency, so that one could see a time when the need for palliative measures under the price and market policy would be reduced or even eliminated. If such a time is not yet in sight, then one needs to question whether the commitment to and content of structural policy within the CAP are adequate.

AGRICULTURAL MARKETING IN DEVELOPING COUNTRIES

Economic reforms have had sweeping impacts on agricultural markets in developing countries.

In general, state intervention has been reduced, notably with respect to:

- The abolition or sharp curtailing of parasitical marketing boards;
- Depreciation of formerly over-valued currencies rendering developing country exports more competitive and imports more expensive;
- A reduced public role in agricultural services, especially in subsidized credit, input and extension networks;
- A shift away from pan-territorial and pan-seasonal crop pricing strategies and pre-announced prices.

Reviewing agricultural markets research in sub-Saharan Africa and Asia, Jones concludes:

- "In newly liberalized [food] markets in eastern and southern Africa... barriers of entry to trade are low, but the marketing system has little capacity to channel credit or spread risk. There are strong theoretical reasons for expecting the impact of and response to reforms to vary between different classes of producers. The absence of key markets, risk aversion, high transaction costs and the dual role of agricultural households as producers and consumers are critical features. The marketing system depends on both physical and institutional infrastructure... Collective action by market participants may address this but it may also lead to collusion over prices. Evidence from South Asia shows that food markets exhibit social barriers to entry, massive asset polarization, debt relationships between large and small traders and traders and farmers, diverse institutional and contractual arrangements, and collusive behaviour, enforced in part by manipulation of the state regulatory system."

This conclusion gives some clue to the reasons why NGOs and CBOs intervene in agricultural markets. When extension agents, researchers and development organizations working in rural areas ask farmers to prioritize their problems, agricultural marketing is repeatedly raised as one of the most important problems faced. It may arise in the context of the promotion of new crops or productivity-enhancing technology, or it may be felt particularly acutely in remoter areas poorly served by commercial traders, where parastatals no longer operate.

NGO marketing interventions typically aim to fill critical gaps in the marketing system or address the power imbalances to which Jones refers. Nowhere are those marketing problems felt more acutely than in the areas for which it is most difficult to identify sustainable strategies to improve market access. Farmers in remote areas (either remote because of physical distance from markets or because of poor roads) are almost always poorly served by agricultural traders and are often obliged to accept seemingly unattractive prices for their produce. Distance from markets rules out the production of higher value more perishable crops, and reduces the linkages between these producers and other more specialized markets. By the same token, CBOs and NGOs seeking to promote alternative strategies for these disadvantaged communities face high costs and tangible obstacles that make their task particularly difficult.

Poor access to markets is mirrored by poor access to all kinds of rural services. The poverty that results makes such communities particularly risk-averse. Where rainfall is uncertain, the situation is even worse, whilst the relative absence of trade does nothing to relieve the covariance in production. These are the challenging circumstances that make an examination of marketing interventions worthwhile. There is wide-ranging experience amongst the development NGO community. Some of these initiatives have taken-off and developed into self-sustaining activities, whilst others, although not conceived as such, have effectively become subsidy dependent welfare programmes. This review identifies best practice and the conditions required for such programmes to work.

NGOS AND CBOS - SOME DEFINITIONS

NGOs are part of the development landscape. Increasing amounts of development aid are channelled through NGOs.

The term gives little clue as to their real characteristics but most people associate NGOs with the following:

- A formal and officially recognized organization that is not linked to government;
- Having a purpose that is altruistic rather than commercial;
- Attracting staff who are value-driven rather than financially motivated.

Fowler highlights some features of the voluntary sector by making comparisons with government and business- organizations.

Table. Comparison of Organisations in Different Sectors

		Sector	
Characteristics	**Government**	**Business**	**Voluntary**
Relationship to those served based on:	Mutual obligation	Financial transaction	Personal commitment
Duration of relationship to those served:	Permanent	Momentary	Temporary
Approach to external environment:	Control and authority	Conditioning and isolation	Negotiation and integration
Resources from:	Citizens	Customers	Donors
Feedback on performance	(in) direct politics	Direct from market indicators	"constructed" from multiple users

The term 'community-based organization' or CBO may be associated with similar values but is generally more focused on issues particular relevant to the community from which its membership is drawn. CBOs may be quite formally structured, but can equally be loosely structured, informal organizations; farmers' associations are an example of a CBO.

Stocker and Barbor-Might discuss CBOs in relation to civil society organizations (CSOs): "CSOs are, simply, organizations operating in civil society somewhere between the informal associational world of family, kin, neighbours, friends and the more formal or market-oriented world of business organizations, state – and NGOs. Although in practice there is diversity and complexity, organizationally the idea is a simple one; NGOs act as intermediaries for CSOs... CSOs sometimes evolve into NGOs. Many of them have nothing to do with development or are not poorer people's organizations. With this in mind, many authors and authorities prefer to use alternative terms, for example, 'GRO' (grass roots organization) or 'CBO' (community-based organizations) to designate CSOs that exist to serve their members, these members being poorer people. CBO is the usage followed by the World Bank."

NGO AND CBO INVOLVEMENT IN AGRICULTURAL

MARKETING

DIFFERENT TYPES OF ORGANIZATION

Although some definitions were provided in the previous section, these were not especially helpful in distinguishing between the plethora of organizations present in many developing countries.

A four-way categorization is proposed here, based largely on origins and capacity:

1. Northern NGOs with offices in developing countries, usually obtaining funds from donors (including private individuals); this group is quite broad since it encompasses large NGOs such as Oxfam or CARE, with activities in many countries, as well as small NGOs whose activities may be quite focused on a few countries and issues.
2. Indigenous NGOs who have become relatively large, well-organized and able to attract significant funds from international donors and Northern NGOs; sometimes these NGOs were originally created or strengthened by Northern NGOs.
3. Indigenous NGOs that are small, usually focused on a particular geographical area or issue, that obtain small amounts of funding from donors or government, but who struggle to grow or stay afloat.
4. CBOs, membership organizations serving particular interest groups usually in rural communities, whose focus may be broad or quite narrow; these organizations may be formally structured or quite informal; farmers associations, credit groups, and joint marketing societies could all be considered CBOs in the context of this review.

Many countries have laws governing the registration of different types of organizations that may confer a certain tax status or legal standing. Some developing countries have umbrella associations for NGOs. In any particular country it is useful to find out whether an umbrella organization exists, and if so, the types of CBOs and NGOs that tend to be officially registered – recognizing the potential divergence between official requirements and practice.

THE EVOLVING ROLE OF NGOS AND CBOS IN DEVELOPMENT ASSISTANCE

In the last 20 years, NGOs have become progressively more involved in development assistance, at every level. The shift from a relief and welfare focus has come about partly in an attempt to address the underlying causes of some of those man-made disasters or to limit the negative consequences of the natural disasters at which they assisted. It has been helped by the increased funding they found they were able to attract. Many Northern NGOs now have policy and research departments, and are a legitimate channel for large amounts of donor funding. At the same time, the role of the state has been redrawn, and in developed and developing countries, there is now a much greater focus on civil society as a way to improve democratic processes and bring about greater accountability in government. Governments are also seeking ways to be smaller and to sub-contract functions where feasible. Furthermore, funding developing country organizations to carry out development work is considered a way to build indigenous capacity. NGOs working in developing countries have benefited from this trend – either because they are considered part of civil society or because they work closely with many civil society organizations, including CBOs. As Stocker and Barbor-Might state:

- "From the point of view of the donors, civil society was the 'place' where something could be done and, often enough, NGOs were the intermediary institutions or midwives of such remedial programmes [relating to structural adjustment], spanning the gap between donors and CSOs. The funding channels varied, sometimes being directed through Northern NGOs (which might provide 'aid' directly or channel it to one or more partner Southern NGOs or CSOs) and sometimes going as direct funding to Southern NGOs and in some instances even to CSOs. When governments were irredeemably corrupt or oppressive (as, for instance, in Haiti during the Duvalier regime), these programmes seemed to offer virtually the only hope of channelling assistance to the people who most needed it."

This growth in the funding, remit, competencies and responsibilities of NGOs has meant that they have been closely involved in (if not the instigators of) much of the experimentation with practical solutions to pressing problems in rural areas. This is the context in which NGO experiences with agricultural marketing interventions provide a valid and rich focus for this review.

NGOS AND CBOS

Although many NGOs share similar altruistic goals, their approaches vary enormously. This is particularly evident in the extent to which they embrace and harness commercial activities to promote broader objectives, or reject this as a legitimate means by which to achieve social objectives. Moreover, amongst those NGOs prepared to use commercial activities as a means to an end, there can be considerable variability in the role these activities are accorded within the development strategy and the competence with which they are planned and undertaken.

Organizations that are Primarily Welfare-oriented

A large number of NGOs and CBOs become involved in agricultural marketing activities, but this is rarely their core business. (There are some notable exceptions amongst some of the international NGOs who have become very experienced in agricultural enterprise, agro-processing and marketing. These include, for instance, TechnoServe, the Intermediate Technology Development Group, Enterprise Works Worldwide and the Cooperative League of the USA.)

Many NGOs start with welfare (or social or altruistic) objectives, in areas such as education, health, water, infrastructure and agriculture and gradually shift towards a longer-term development focus. With this shift, small business and income-generation activities take on a greater role. Gibson describes this gradual transformation in terms of a continuum of different actions and attitudes.

Table. NGOs' Evolutionary Path in the Development of small Businesses and Income Generation

From ←	→ To
Relief and welfare	Development
Short-term	Long-term
Ideological	Pragmatic
Community-focused	Individual-focused
Targeted	Self-selecting
Grants	Market interest rates
Amateurish	Professional
Income generation	Small business
Social/ Technical	Economic/ Business
Instinctive	Strategic
Beneficiaries	Clients

Often NGOs and CBOs deliberately work in remote and disadvantaged communities and target the poorest households or individuals. These conditions, in combination with a general relief and welfare orientation, influence the strategies they adopt to achieve their objectives. For instance, direct or indirect subsidies may be used to improve access to markets (*e.g.* through the provision of transport, credit or inputs). An example of a direct subsidy is free or below cost use of transport (calculated on the basis of costs of fuel and driver and perhaps some portion of the vehicle costs). An indirect subsidy might involve charging a commercial (or break-even) rate on the vehicle hire but taking no account of the staff costs of implementing and managing the scheme. Whilst few people would suggest that the subsidy could continue indefinitely, there may be little consideration of how these activities can eventually be shifted to a more sustainable basis. The result is often that the programme attracts participation because of the subsidies, and once it ends there is little enduring impact.

Yet in the short-run these types of activities are attractive to NGOs because they have fairly immediate and visible (if not enduring) impacts and can (with varying degrees of success) be targeted to particularly disadvantaged groups (such as the poorest households, women, refugees, the handicapped or other marginalized social groups). An approach that seeks to use commercial channels may take much longer to develop and may place the intended target group at a disadvantage relative to other members of the community. Even when NGOs and CBOs do not intentionally adopt a non-commercial approach, market-oriented interventions are often subordinate to their core business.

This affects the way they are developed and managed, as well as the way they are perceived within and outside the organization.

Stanton gives four reasons for this lower status:

- Ideological – "...the core work assists those who most need it, income-generating work assists those who will exploit it to the greatest economic advantage";
- Resource allocation – reflecting the ideological perspective;
- Management – "the most willing volunteer for the job rather than external recruitment of people qualified and experienced in business management";
- Finance – these activities may be quite costly to implement and effectively monitor.

Furthermore, marketing activities are often managed and evaluated in the same way as other development activities, with insufficient attention to budgeting and profitability. A survey of income-generating programmes carried out by a range of indigenous and international NGOs in Welaita, south-west Ethiopia, found that little attention was given to the income generated and the profitability of different schemes. However, NGOs using subsidies to target disadvantaged groups would argue that this is a legitimate way to improve the livelihoods of poor individuals, households and communities, particularly in remote areas. Yet even with these subsidies, it may be difficult to have much impact on livelihoods in the most geographically and socio-economically disadvantaged communities. Stanton points out the disadvantages of this type of approach: the frequent failure to make a significant profit, high costs that prevent the NGOs/CBOs from reaching a wide audience, and problems concerning long-term sustainability.

Business-like NGOs and CBOs

- "Organizations which themselves resemble small businesses – in terms of their people, culture, systems, structure and behaviour – are most likely to be successful in encouraging the growth of small businesses."

In recent years, private sector development has increasingly been seen as a viable and important approach to sustainable development. Thus many governments, NGOs and CBOs have focused on the promotion of marketing and small-scale enterprise to encourage greater participation in the commercial sector, as a route to higher incomes, employment generation and growth.

Small enterprise development work, which grew considerably in the 1980s, has contributed to a realization that it is possible to make much greater use of market mechanisms in pursuit of development objectives. This has been shown particularly through the success of micro-credit initiatives, where even very poor individuals are able to repay not only loans but also interest which sometimes covers the costs of providing credit. Also, social objectives and commercial objectives are not mutually exclusive and many NGOs and CBOs pursue both. The fair-trade movement is a good example of this. Fair-trade organizations use commercial methods to generate social development benefits through improved terms of trade. The important thing is to balance potential marketing success with the social benefit needs of the beneficiaries. Furthermore, selective use of subsidies can still lead to sustainable and successful marketing initiatives, depending on the circumstances, as demonstrated by the CARE Egypt Agricultural Reform Programme. The programme provides information services to smallholder farmers and facilitates linkages to help increase farmer income.

The service is highly subsidized but has proven successful for a number of reasons:

- It helps to link farmers to sources of information outside the programme, thereby fostering the long-term sustainability of relationships and networks;
- Farmers contribute financially, *i.e.* they pay fees for the services;
- The demand for services is farmer-driven and project staff work with farmers to identify production and marketing opportunities.

A similar approach is used by Intermediate Technology in Zimbabwe. Assistance in product development is offered but the initiative must come from an existing business, which must be willing to contribute to the costs of product development (*e.g.* through materials, labour or workshop facilities). The two key advantages of greater commercial orientation and awareness are that cost-recovery enables more people to be reached by such programmes, and sustainability becomes a realistic goal. Gibson argues that NGOs appear to have distinct advantages in pursuing income-generation programmes ("smaller, more flexible, innovative organizations"). His conclusions are firmly rooted in the belief that commercial strategies can serve development objectives:

- "The continuing challenge is to progress from this base so that the economic growth of other developing countries is enhanced, is driven by indigenously owned and indigenously managed enterprises, and reaches the poor and disadvantaged sections of the population."

DIRECT INTERVENTION OR FACILITATION

The marketing role that NGOs and CBOs take on lies somewhere along a continuum between being directly responsible for marketing activities to facilitating beneficiaries/clients to market for themselves.

Direct Marketing Role

The term 'income-generating programme' (IGP) is used to describe a variety of programmes. These range from enterprises owned and managed by the beneficiaries to enterprises owned and managed by the organization which employs the beneficiaries. A number of NGOs/CBOs have established this latter type of small business to generate income to finance their other programmes and reduce donor dependence. CBOs and NGOs can also be more directly responsible for marketing activities.

One way of doing this is through out-grower schemes (sometimes referred to as contract farming or satellite production). Such schemes involve smallholder producers providing agricultural raw materials to trading or processing businesses. Often growers work as a group, characterized by String fellow *et al.* as linkage-dependent groups. Generally the marketing arrangements are predetermined: prices or a pricing formula are agreed. They help markets function to the benefit of both producers and the companies or organizations involved.

This type of relationship is beneficial for farmers because they have a secure market for their produce at a predetermined price and the buyer benefits from having a guaranteed source of raw materials and lower transaction costs, which reduces his/her risk and costs. Within the fair-trade arena, the role of, and the marketing channels used by NGOs and CBOs (or alternative trading organizations – ATOs), also varies. Some organizations (such as Oxfam Trading and Traidcraft) take a direct marketing role by acting as wholesalers, with the producers acting as subcontractors producing to order. An advantage of this type of arrangement for producers is that they are guaranteed a volume of sales, thereby minimizing their risk. A disadvantage of this, and of out-grower schemes, is that the producers can be dependent on the trader, and may not have access to alternative buyers or markets if for any reason the trader is no longer able to market their produce.

ADC and Bean Outgrower Schemes in Uganda

The Agri-business Development Centre (ADC) in Uganda helped to establish bean outgrower schemes in Kasese and Kibaale districts together with the Uganda National Farmers Association (UNFA) and Bugangaizi Export Commodities Limited (BEC). The purpose of the schemes was to integrate poor rural farmers into the bean market, through marketing agencies, to provide them with an additional income source. ADC's role was facilitating the supply of bean seed and training some of the producers as 'farmer-extensionists' to provide follow-up extension advice to other producers. UNFA and BEC promoted and implemented the programmes. They were responsible for managing seed supply, training and extension and organized marketing (procurement and collection) of the crop.

The outgrower schemes have been successful from the producers' perspective in several ways:

- The numbers of farmers reached has steadily increased;
- Output levels have increased (participation, acreage and yields have all increased);
- An effective farmer-extension system has been created, and the

adoption of new varieties and improved production methods has been good;

- Planting materials have been maintained and expanded through seed multiplication;
- Household income from bean sales has increased.

From the trader perspective, a strong linkage was formed with a private buyer in Kasese, who procured beans through UNFA and sold to a Kampala-based exporter. In Kibaale the linkage was weaker, due to the inability of BEC to raise finance. This lack of capital was a key constraint, and undermined the efforts that had gone into developing the scheme and building effective outgrower loyalty. Also, competition from other traders emerged for the beans, which affected potential profit. This highlighted the importance of product selection within outgrower schemes and the need to consider diversion factors (whether the product can be used or sold outside the outgrower scheme) and the buyer's (financial) exposure ratio (the likely cost of obtaining the crop against anticipated sales value).

Facilitative Role

Other NGOs and CBOs play a more facilitative role. They assist individuals, groups and communities to market for themselves. This includes both improving access to, and benefits generated from, existing products and existing markets as well as creating new products and new markets (*e.g.* through technology development and processing). There are a variety of ways in which organizations facilitate marketing, including: strengthening the capacity of individuals, groups or communities (through group strengthening and training); developing linkages to traders and other stakeholders in the marketing chain (*e.g.* input suppliers, credit sources and transport agents); and linking farmers to relevant market information. This type of facilitative role is beneficial for a number of reasons: being less interventionist, it is likely to generate more sustainable marketing activities and linkages; it is likely to be achieved at lower cost than if the NGO was more directly responsible for marketing activities; and, therefore, it facilitates reaching a wider audience.

5

Farm Management System

Farm management deals with the organisation and operation of a farm with the objective of maximizing profits from the farm business on a continuing basis. The farmer needs to adjust his farm organisation from year to year to keep abreast of changes in methods, price variability and resources available to him. Thus farm management is the science which deals with the analysis of the farming resources, alternatives, choices and opportunities within the framework of resource restrictions and social and personal constraints of farming business. This complex information is integrated and synthesized to increase profitability of the farming business, the ultimate aim being to raise the standard of living of the farming people. This does not mean that farm management deals exclusively with the maximization of income; in fact, it takes into account the goals and objectives of the individual farmer, other than income maximization. Thus this discipline deals with people or organisers and decision-makers in respect of farms and agricultural production. It is people-oriented rather than crops or livestock per se. Farm management is a decision-making science. It helps to decide about the basic course of action of the farming business.

The basic decisions of the farming business are:

- What to produce or what combination of different enterprises to follow?
- How much to produce and what is the most profitable level of production?
- What should be the size of an individual enterprise, which, in turn, will determine the best overall size of the farm business?
- What methods of production (production practices or what type of quality of inputs and their combination) should be used?
- What and where to market?

CHARACTERISTICS OF FARM BUSINESS SYSTEMS

The 31 percent of farms with the least complex farm business system controlled 11 percent of acres operated and generated less than 3 percent of production value in 2003. Over 85 percent of these farms are rural residences and the rest almost entirely intermediate farms (sales below $250,000 and the operator reports farming as his or her major occupation). Over 98 percent had sales of less than $100,000. These farms specialized in production of field crops other than cash grains or soybeans, beef cattle and general livestock. Operator and operator-spouses managed three-fourths of the least complex farms. They had the highest average operator age and the largest share of primary operators over age 65 years (32 percent). Over 80 percent considered their primary occupation to be off-farm and almost 29 percent were retired. This helps explain the 890 hours worked onfarm by the operator, well below the all-farm average of 1,393 hours.

The most complex farms controlled 8 percent of acres and generated 18 percent of value of production in 2003. Farms in this group were mostly commercial. Over 27 percent had over $500,000 or more in sales (versus 3 percent of all farms). The most complex farms had a much larger share of management teams that included multiple persons. These farms were more common in the Northern Crescent, Heartland, Northern Plains, and Eastern Uplands. They tend to specialize in dairy, poultry, hogs, cash grains, and soybeans. Nearly two-thirds reported hiring other individuals and three-fourths had custom hire assistance. The primary operators in these management units were younger, averaging 49 years, nearly 7 years less than the all-farm average. On average, primary operators reported working over 3,000 hours on their farms in 2003, with spouses and other operator labour adding more than 1,100 hours to the total.

MANAGEMENT OF A FARM

Farm management is different from what is commonly confused with the work of a farm manager who manages a government farm as an agronomist. His function is normally limited to supervising and handling the day-to-day routine of a farm. It normally pertains to the existing pattern of the resource-use and crops-mix under which only the existing plan is executed, supervised and carried out. An intelligent farm manager may go a little further and look after the farm machinery to keep it going.

Farm management, is however, much more than that. Here we are not just concerned with the distribution of labour and irrigation water for day-to-day operations. The emphasis is on the decision-making function of evaluating and choosing between alternative strategies. A major concern is about adjustments which are more suitable and profitable and about exploring new situations and opportunities for maximization of income and satisfying other goals of a farmer. It is the approach under which the opportunity costs of the various resources are evaluated and adjustments in resource-use and enterprise mix are made to secure higher levels of farm income.

Following this approach, the emphasis on yield (productivity per unit of a resource, mainly land) is not ignored, but the greater focus is on the increasing of farm income through a second business organisation. As a business, farming requires the application of business methods and efficient management. To be an efficient manager, one must keep oneself abreast of developments in new technology, new practices, price trends, and economic outlook. Again a farm management man should identify the constraints in the external environment conditions, which hamper a farmer's opportunities and plans for making adjustments in his farm organisation and render the technically superior production plan economically unattractive to the farmer.

MEANING OF FARM MANAGEMENT

Farm Management is the application of business principles and the scientific principles of agriculture (as discovered by the chemist, the physicist, the agronomist, the animal husbandmand and other specialists) to the business of farming. The farm manager is the one who plans the farm, arranges the cropping scheme, provides the equipment and actually transacts the business and sees to the operation of the farm. The term is a new one and it is important that students of farming get a clear understanding of its meaning. Because of the close relation of farm management to soils, farm crops, live stock, and other agricultural subjects, people often confuse the terms and gain erroneous ideas of what farm management really is.

Farm management is also related closely to rural economy. There is danger of confusion arising in the consideration of these terms also. Questions which relate to the single farm as a unit or to a group of farms operated under the same organization are matters of farm mangagement. Those which relate to a group of farms operated separately, or to a community, are matters of rural economy. The farmer must meet both classes of questions. To state it differently, those things that have to do with the organization and operation within the individual farm relate to farm management. Those that have to do with the external relations of the individual farm to the group or community, relate to rural economy.

The Relation of Farm Crops to Farm Management

The farmer must grow crops on his land either to sell or to feed to his live stock. This makes it necessary for him to know how to prepare the land, sow the seed, cultivate the crops and harvest and store the product. In gaining this knowledge he studies the questions from the viewpoint of the agronomist and considers only what will be necessary to provide the best conditions for the growing crop.

The farmer must also decide what crops he will grow. One farmer will want to grow corn, another will grow cotton, and still another will grow hay, wheat or potatoes. Usually he will grow several different crops. He will need to decide how much of each one should be grown. In deciding these questions he will study the questions from the viewpoint of the farm manager.

As an agronomist, the farmer is interested only in how to grow the crop. When the crop is mature and harvested, the problem is completed. As a farm manager he still has to settle the question of what to do with the crop. It may be sold. Would it pay better to feed it to live stock? If so, what kind of live stock will give the best returns for the feed used? These are farm management questions.

Relation of Animal Husbandry to Farm Management

Most farmers grow some live stock. They must know the nature of the animals they wish to grow, the kind of food they eat, how to prepare it for them and when to feed them. They should also know how to select the best animals and to what purpose the various classes of animals are best adapted. In answering these questions they study them from the viewpoint of the animal husbandman and cosider only what is best for the animals.

The farmer must decide, however, what classes of stock he will grow. On some farms dairy cows will be best, on others beef cattle and hogs will be a desireable combination. There are farms on which horses and sheep with some poultry will be preferable. There are also the question of how much live stock the farm will supply feed for, or what to do with the surplus when more is produced than is required for the live stock. In answering these questions they are studied from the view point of farm management.

Relation of Rural Economy to Farm Management

The management of a farm includes the buying and selling of products. It is important that the farmer have a knowledge of markets and of values, and that he know when to buy and sell. Often cooperative marketing offers the best opportunity for disposing of products at a profit. So long as the farmer is concerned only with the business enterprises relating to his own farm, he is within the field of farm management. When the business activities and social welfare of several farms or an entire community are collectively considered, the question becomes one of rural economy. Farm management deals with the problems of the individual farm from the business point of view. Rural economy deals with the problems of a collection of farms or an entire community.

Studying Farm Organization

The student of farming is taught how to grow crops by the agronomist, how to grow live stock by the animal husbandman. He learns how to know insects and combat them from the entomologist, and how to cooperate with his neighbors in selling his crops or his stock, from the rural economist. Finally, from the viewpoint of a farmer, he will need to select a specific farm, make plans for its arrangement into fields, select the kind of stock and crops that it will be best to raise, and determine how much of each should be raised. He must provide the buildings and equipment, estimate the capital required and the labour necessary to operate the farm. The farm is thus organized into business form and its operation is simplified. This art of farm organization is learned from the farm manager.

Meaning Made Clear by Questions. "How shall I grow corn?" is clearly and agronomy question. How to select and grow cattle, is just as clearly a question for the animal husbandman. "Shall I grow corn? or some other crop? or cattle? Having grown corn, shall I sell it? or feed it to cattle? or hogs?" are farm management questions. If questions similar to the above are put to each problem presented there should be no difficulty in distinguishing the viewpoint from which it should be attacked.

Definition of Terms. In the pages following, reference will be made to the terms, "Capital," "Inventory," "Farm Receipts," "Farm Expenditures," "Farm Income," and "Labour Income." An explanation of these terms is given below.

Capital. Economists divide capital into two classes: (1) Fixed capital, including land, buildings and permanent equipment, as teams, implements and live stock for purposes of production; and (2) circulating capital, which includes values in seed, feed, fertilizers, supplies and market stock or crops as well as cash for operating the farm. The term is limited to those forms which disappear in the process of production, reappearing in other materials of similar kind.

There is a tendency on the part of farmers to tie up most of their capital in fixed forms, leaving too little for circulating or operating capital. The Inventory is a detailed list of the farm property with values assigned. An average of the values at the beginning of the year and at the close is considered to be the value of the capital invested in the business. The difference between the opening and closing inventories shows the increase or decrease in the value of the farm property. Farm Receipts are derived from two sources: (1) direct and (2) indirect. The direct receipts are represented in the cash received from the sales of crops, live stock, live stock products, for labour for others and from miscellaneous sources. The indirect receipts would arise in an increase in the inventory due to permanent improvements or to appreciated values of any of the various forms of property. Values should not be inflated for the purpose of showing a gain in the inventory. Debts cannot be paid on anticipated rises in land values or in values of property unsold.

Farm Expenditures

Items paid for in cash for the support of the farm only can properly be classed as farm expenditures. Items of personal or family expense should not be charged against the farm in determining the profits from farming. A decrease in the inventory is considered as an indirect expenditure. The amount of the decrease should be deducted, with the direct expenditures, from the receipts to show the net farm income. The Farm Income is the difference between receipts and expenses. It is the return for the use of capital and unpaid labour. The farmer and his investment unite in earning the farm income. To learn what is earned by the unpaid labour only, the amount that the value of the investment would earn if placed at interest, must be subtracted from the farm income.

The Labour Income is the amount earned by the farmer alone. To learn what this amount is, the amount earned by the money invested at the local interest rate and the value of the labour of other members of the family at local wages while employed at labour on the farm, must be deducted from the farm income. The balance will be the labour income of the farmer.

FARM MANAGEMENT AND OTHER SCIENCES

Basically farm management is economics in the context of fundamental definition of economics which involves three elements, *viz.* the scarcity of resources, their alternative uses and the objective of profit maximization. It is this science which deals with making rational, profitable and economic recommendations to the farmers. It also deals with the growth and stability aspects of economics.

The basic information about the multiple (alternative) uses to which the scarce resources can be allocated is supplied by other physical and biological sciences. The research in these sciences should continually generate the relevant data on alternative technologies and practices whose profitability can be tested under actual farm situations. And these data are required for the farm organisation as a whole and not far for a single hectare, animal or tree. The challenge to the farm management specialist is to be able to integrate and synthesize the diverse pieces of information from many disciplines such as agronomy, animal husbandry, horticulture, soil science, plant breeding, entomology, plant pathology, general economics, sociology and psychology, into an optimum 'package' which can be used on the farms with profit.

Agricultural scientists mainly put emphasis on the maximisation of yield rather than on the use of the optimum level of resources. But the goal in farming is not to make a profit on some single enterprise or from a part of the farm land as some *krishi pandits* do, but to use land, labour, and capital resources in such a way that they make the greatest contribution to the total profits from the entire farm. The superiority of this discipline, thus, lies in its treating the farm as an operational unit and tailoring the recommendations of all other disciplines to fit into an individual farmer's pattern of resources. In the context of the recent technological breakthrough, management today should be viewed as a process within a rapidly moving frame of reference. "It is now more scientific, less artistic; more dynamic, less static; more versatile and less rigid". Farm management is forward-looking in its approach. Its task is not so much the improvement of the present farming practices but of the establishment of the whole sets of new production methods and farming systems which would put our agriculture on a continuously rising growth curve.

Government policies should change the economic environment to help the interests of the farmers to converge on the national goals. Studies on farm management to determine the responsiveness of the farmers to different levels of prices become extremely relevant in inducing the farmers to produce the quantities of different agricultural products and services needed by society. A rational pricing policy for water is needed. Canal water, for example is charged for without any direct relation to the quantity supplied. As a result, there is no incentive to the farmer to allocate water in the most economical manner. Farm management helps to identify such uneconomic practices and the most limiting factors. Irrigation or power (the bullock versus the tractor) may be a more important restriction than the limits imposed by land. In areas where water rather than land is the principal limiting factor and the marginal-value productivity of irrigation water is very high, it should be most carefully used rather than over irrigating a few hectares of crops needing high water consumption.

Good farm management can lead to a highly productive use of farm resources and can avail itself of the technological revolution now going on in our agriculture.

APPLICATION OF FARM MANAGEMENT

As against the general contention that farm management is mostly applicable to crop farms, its application to other types of farming business, such as orchards, dairying, poultry, etc. is no less important. Faculty farm management has contributed more to citrus decline than the poor physical-chemical conditions of the soil, the stock-scion incompatibility, insect pests, disease and viruses. Intercropping with exhaustive crops seems to be one of the major reasons for the poor health of the plants. Citrus die-back or decline does not seem to be a disease by itself, but a mere symptomatic expression of physiological imbalance owing to the cumulative effect of several factors. Similarly, if milk production has to be encouraged, it has to be made more profitable than cereal production. One has to demonstrate how the addition of milch animals to a farm fits into its resource organisation and yields higher returns.

With land as a scarce resource, efforts are made to intensify its use for maximising profits through multiple cropping. During a given period, the relative profitability of different crop rotations is evaluated and finally selected for adoption. 'Kufri Sinduri' variety of potato is not being adopted on a large scale in spite of its giving 10 per cent higher yield than some other varieties, *e.g.* 'Up-to-date'. The reason is simple. The former is a late variety and delays the sowing of wheat which decreases its yield. It is not 'Kufri Sinduri' potatoes versus 'Up-to-date' potatoes; rather, it is 'Kufri-Sinduri' followed by wheat versus 'Up-to-date' followed by wheat that counts in farm management. Similarly, in West Bengal, where jute has been competing with paddy, the situation is changing with the introduction of high-yielding varieties of wheat, which can be rotated with paddy. The comparison in terms of the relative profitability now runs between jute and paddy plus wheat, and not just between jute and paddy. Owing to the paddy-wheat rotation becoming more profitable, jute production in West Bengal is suffering a set back. As Indian agriculture becomes more productive with the passage of time, the farmers will be confronted with the task of making the best selection from among many alternatives.

SCOPE OF FARM MANAGEMENT ON SMALL FARMS

Sometimes it is said that management is not important on small farms. This notion is widely prevalent under those situations where the farms operate on a low level of technology. In the case of small farms, farm organisation cannot be changed as much as on a large farm, but the choice in respect of farm practices and methods of production, cropping intensity, etc. offer worthwhile alternatives. Japan's method of improving agriculture based on the principles of farm management clearly demonstrates that a small farm by itself should not be a hindrance to increased production and higher income. Small farms may have different problems, but the principles of the efficient use of available resources to obtain the maximum economic returns and family satisfaction remain the same in both the cases. The principles of farm management, since they deal with the allocation of resources, apply to small and large farms equally. A Chinese proverb is very appropriate to quote: "Although the size of a sparrow is small, yet it is physiologically as perfect as a big bird".

MEASURING THE SIZE OF BUSINESS

Farm income is directly related to the size of the business but in no way it indicated whether the farming business is operating efficiently or not. The following are some important measures of the size of a business.

Total Area of Farm and the Number of Livestock

The acreage as a measure of the size of the farm may be either the total land, or the area under crops. The acreage at its face-value carries no meaning. It is necessary, therefore that land be homogenised and standardised for variations in fertility status, irrigation and lack of irrigation, etc. by such standard measures as rental value, land revenue, etc. As in a given type of farming area, there are important crops common to all farmers; therefore, the area (standardised) under principal crop enterprises will help to introduce further refinements in this measure.

In the case of livestock farms, the number of animals can be used as a measure of the size of the business, again, the breed differences between the herds of two farms make a great difference in the returns received by them. On farms having both crop and livestock enterprises, any of the two will not give the relevant size of the business. Both may be used to represent the situation, but it is not comparable.

Total Capital Managed

The total capital managed is obtained by adding all the capital values of inventories, including land either of the closing and beginning inventories can be used. This measure is useful when all the farms are producing the same product and this has no meaning for different specialisations. One is also not sure at the face value of the capital investment whether it is nearer to the optimum required, one farm can be compared with the other for the size of business by the capital investment only if the two farms have the near-optimum investments. The difficulty here is that many farms are under stocked or overstocked and the comparison between them is meaningless.

Total Inputs

The level of inputs in a broad way determines the level of output income. The total inputs mean the operational expenses plus the fixed expenses- the rental value of land, the value of labour (of family and operator) and the interest on the capital. This is the most reasonable method. It can be used in similar situations to compare with the assumption that rates of rent, interest and wages would be the same for each comparable situation.

Gross Income

This relates to the farm output during the year. This is obtained by adding the home-consumed farm products (in ₹) and the increase or decrease in inventory to the gross receipts of the year. This is a good measure but its use is limited to the same type of farming.

Productive Manpower Units

They refer to the total manpower input required for normal efficiency. This method is more appropriate where labour is scarce and is being used optimally. In a labour-surplus situation, this method has only limited value. Also, the measure is limited to the comparing of the similar type of farming situations. Again, the difference arises from the difficulty to see whether the farmers under comparison use the scarce labour where it adds to the maximum or not. These measures are rough indicators of the size of the business and can be used to compare only similar situations at a point of time and for the same type of farming.

Cost Concepts

There are many cost concepts which are in frequent use in the farm management literature, *i.e.* Cost A_1, Cost A_2, Cost B and Cost C; working or variable or operational cost and fixed costs; machine costs, labour costs, livestock costs, crop costs, etc. a brief discussion of these cost concepts is given below:

i. *Cost A_1*:

It includes the following 16 items of costs:

1. Value of hired human labour (permanent and casual)
2. Value of owned bullock labour
3. Value of hired bullock labour

4. Value of owned machinery
5. Hired machinery charges
6. Value of fertilizers
7. Value of manure (produced on the farm, and purchased)
8. Value of seed (both farm produced and purchased)
9. Value of insecticides and fungicides
10. Irrigation charges (owned and hired tube-wells, pumping sets etc.)
11. Canal-water charges
12. Land revenue, cesses and other taxes
13. Depreciation on farm implements (both bullock-drawn and worked with human labour)
14. Depreciation on farm buildings, farm machinery and irrigation structures.
15. Interest on the working capital
16. Miscellaneous expenses (wages of artisans, cost of ropes and repairs to small farm implements)

ii. *Cost A_2*: It is cost A_1 plus
   1. Rent paid for leased land

iii. *Cost B*: It is cost A_2 plus
   1. Imputed rental value of owned land (less land revenue paid thereupon)
   2. Imputed interest on owned fixed capital (excluding land)

iv. *Cost C*: It is cost B plus
   1. Imputed value of family labour

It can be seen that cost C is a very comprehensive concept. But the risk and uncertainty costs are not taken care of.

- *Variable costs*: These are the costs which are of the recurring type, and have to be incurred during every production period, *e.g.* seed fertilizers and insecticides. These are also known as operational costs or working costs.
- *Fixed costs:* These costs are of non-recurring nature including cost of the tractor and other machinery buildings, irrigation structures, livestock, etc.
- *Crops costs:* These costs refer to the value of seeds and plants, manures and fertilizers, insecticides and fungicides and irrigation charges.
- *Machinery costs:* These costs include the hiring charges of machinery, cost of fuel and lubricants, electricity bills, minor machinery repairs, depreciation and interest on machines and equipment.
- *Labour costs:* They include the wages paid in cash or kind to the hired labour and the imputed value of family labour used on the farm.

- *Livestock costs:* They include the costs of veterinary medicines and services, fodders, feeds, interest on the values of livestock and depreciation.
- *Land costs:* They refer to the rent of land owned and taken on lease.
- *Building costs:* They include the interest on the value of the building structure and depreciation on them.

FARM PLANNING

Planning means taking decisions in advance. It stimulates thinking, broadens understanding and challenges the farmer to move forward. It is a forward-looking approach.

The farm plan helps a farmer to decide how to combine new ideas and old ones to his best advantage. By identifying his credit and supply needs, the farm plan helps him to arrange for the timely supplies of credit, seeds, fertilizers, etc. A specific farm plan setting fort his expected output, expenses and income, serves as a sound basis on which a credit institution can give him production credit, based on his productive capability rather than on his net financial assets. It is out of his income and not through the sale of assets that the cultivator has to pay off his loan. Thus the farm plan or the budget is to the farmer what the blue-print of the architect is to a building contractor. It shows what is to be done and how it is to be done. It furnishes an organised and logical approach to his problems and helps him to work out the solution.

WHERE PARTIAL BUDGETING IS USED

Partial budgeting maybe used where the change in the activity under study would not affect the farm organisation vitally.

It has wide application and can be used to answer a wide variety of questions such as:

- The economic combination of CAN and superphosphate versus diammonium phosphate with a basal dose of CAN,
- The mechanical thinning of fruits like plums and grapes versus chemical thinning,
- The substituting of a more economical dairy ration for a costlier one (a single or two nutrients), and
- American cotton versus hybrid maize versus desi cotton, etc.

Partial Budgeting Distinguished from Full Budgeting

1. Partial budgeting considers a few alternatives which do not affect the organisation vitally, but full budgeting takes care of all the alternatives.
2. In full budgeting, the inventory of the farm, the resource structure, the existing resource use and such problems as overstocking or under stocking of resources, etc. are considered, but in partial budgeting information with respect to a few alternatives is considered.

3. Partial budgeting does not indicate the break-even point as to when to start one practice and abandon another, but full budgeting does.

The best strategy is to make an effective use of the partial budget at different stages of full budgeting.

Limitations of Partial Budgeting

The partial budgeting analysis suffers from some limitations, *viz.* it does not consider all the alternatives open to a farmer within the restraints of his present resources. It only considers a few alternatives and these too, should not affect the farm organisation to a great extent.

Partial budgeting does not always provide a complete solution. As one may demonstrate that mechanical threshing costs less per quintal than the traditional *phalla* method, but this would not serve the purpose. The other questions such as the break-even point of the thresher, *i.e.* the minimum volume of business (*i.e.* 1,000 quintals of wheat to thresh) the possibilities of custom work, the capital available to purchase the thresher, etc. should also be considered. Similarly, the question whether to increase the number of dairy animals may be answered by using partial budgeting. But, in this case the partial budgeting analysis will approach very closely to a change in the complete farm organisation. Again, the partial budget may show American cotton to be less profitable than hybrid maize, but when full budget is prepared, both the activities may not enter the farm organisation.

Like any other farm-management analysis, the result of partial budgets are subject to variations in the input-output prices and field conditions such as soil type, soil fertility, etc. Also the restraints pertinent to a particular farm organisation necessitate the tailoring of these results to ensure their applicability to means for making profits.

Complete Planning and Budgeting

In preparing complete plans for the farm, all the sophisticated analysis of studying an individual cultivator's opportunities, constraints and problems is done. Every little aspect of the farm organisation is examined and then suitable adjustments are studied and a suitable combination as fits in rationally with that given farm organisation is suggested. Various recommendations in terms of augmenting resources, if the need may arise, are also made. It implies the following steps.

1. *Farm Map*: The farm is carefully mapped out, giving its salient features, like soil type soil-fertility and rotations followed. Low-lying areas or other such features are also shown in the map. Then based on the previous crop history, land-capability classification is done and is also shown in the map.
2. *Inventory of Farm Resources*: Every asset on the farm, ranging from hand-tools to sources of power, etc. are inventoried. It does not provide us with the picture of the resources as owned by a farmer,

but we can also work out their use-patterns and their condition, *i.e.* whether they will have to be replaced or whether they will be sufficient for the new plan or some augmentation will be needed.

3. *Examining the Existing Organisation*: Having prepared an inventory of the existing resources and their availability, what we are interested in is their use-pattern within the framework of the existing crop mix, whether the resources are under stocked or overstocked. A careful analysis of the restrictions and weaknesses of the farm organisation is made. The weaknesses maybe many, such as:
   i. Lumpy units may not be fully utilised.
   ii. Irrigation may not be adequate.
   iii. The peak-season labour requirements may not be met.
   iv. Short-term capital may be inadequate.

   It is not simply the weaknesses that are emphasised but the good points of the existing organisation are also to be highlighted. The existing organisation may be improved or new good items maybe adopted.
4. Laying Down Restrictions and Planning.
   (a) Restrictions maybe with regard to bullock-power availability, or in respect of putting some area under cotton, or vegetables for home-consumption, though such a change in the cropping pattern may not be profitable.
   (b) Labour-requirements, power requirements, capital needs and new equipment needed are worked out and a suitable crop mix is adopted.

In complete farm planning, the goals of the farmer are also important. He maybe desirous of investing in farming and, at the same time, he maybe willing to construct a new house. His goals and opinions and risks are all studied and weighed, and a compromising decision maybe taken. Thus in a complete farm-planning, due consideration is payed to resource use, restrictions, importance of relationships among different enterprises and to the goals and managerial skills of the operator.

RESOURCES OF RURAL FARMING

In spite of propaganda raised by organizations promoting vegeterianism and in spite of threat of Salmonella and cholesterol, world poultry industry is expanding, as the population is increasing. Per capita consumption is also increasing. Not only that, many countries which are not traditionally poultry-growers are giving incentives to their poultry industry. Indian government is also giving incentives to this industry by giving incentives to small poultry farmers as well as poultry industry in organized sector as poultry industry generates employment and also provides proteins to masses. Many countries have to import feed for their poultry industry. India is lucky. We export 24 million tones of soyabean cake to Europe, which is the main ingredient of poultry feed. We can export eggs or chicken meat instead of soyabean-cake.

FERTILIZER IN ORGANIC FARMING

Fertilizer is any organic or inorganic material of natural or synthetic origin that is added to a soil to supply one or more plant nutrients essential to the growth of plants. A recent assessment found that about 40 to 60 per cent of crop yields are attributable to commercial fertilizer use. They are essential for high-yield harvest: European fertilizer market is expected to grow to €15.3 billion by 2018. Mined inorganic fertilizers have been used for many centuries, whereas chemically synthesized inorganic fertilizers were only widely developed during the Industrial Revolution.

Increased understanding and use of fertilizers were important parts of the pre-industrial British Agricultural Revolution and the industrial Green Revolution of the 20th century. Inorganic fertilizer use has also significantly supported global population growth—it has been estimated that almost half the people on the Earth are currently fed as a result of synthetic nitrogen fertilizer use.

Fertilizers typically provide, in varying proportions:

- *Six Macronutrients:*
  - Nitrogen (N),
  - Phosphorus (P),
  - Potassium (K),
  - Calcium (Ca),
  - Magnesium (Mg), and
  - Sulfur (S).
- *Seven Micronutrients:*
  - Boron (B),
  - Chlorine (Cl),
  - Copper (Cu),
  - Iron (Fe),
  - Manganese (Mn),
  - Molybdenum (Mo), and
  - Zinc (Zn).

The macronutrients are consumed in larger quantities and are present in plant tissue in quantities from 0.15 per cent to 6.0 per cent on a dry matter (0 per cent moisture) basis (DM). Micronutrients are consumed in smaller quantities and are present in plant tissue on the order of parts per million (ppm), ranging from 0.15 to 400 ppm DM, or less than 0.04 per cent DM. Only three other macronutrients are required by all plants: carbon, hydrogen, and oxygen. These nutrients are supplied by water and carbon dioxide. The nitrogen-rich fertilizer ammonium nitrate is also used as an oxidizing agent in improvised explosive devices, sometimes called *fertilizer bombs*, leading to sale regulations.

Labelling

The labelling of fertilizers varies. In most countries the macronutrients are labeled with an *NPK* analysis. The three numbers on the fertilizer label represent an analysis of the composition by weight. These three numbers correspond to nitrogen, phosphorus, and potassium and always appear in that specific order.

When a 4th number is included, it indicates the sulfur content.While the number for "N" represents the percentage weight of nitrogen, in some European countries, the other two components are not for the analysis of the element, but rather, the analysis of the "available" or "soluble" form of the element. In traditional chemical analysis, the tests used treated the sample so as to measure the equivalent P_2O_5 and K_2O. For instance, some potassium-bearing rocks do not count as having available potassium.

Similarly the number for "P" in some countries is actually the weight of an equivalent quantity of P_2O_5 whereas in others it refers to elemental phosphorus. In order to calculate the weight of P in the formulation, the weight of P_2O_5 can be multiplied by 0.44 to compensate for the weight of the oxygen in the molecule. For example, a bag of 10-10-10 has 10 pounds of nitrogen, 10 pounds of P_2O_5, but only 4.4 pounds of P. Likewise, the number for "K" in Europe can refer to the weight of an equivalent quantity of K_2O, whereas in Australia it refers to elemental potassium.

In order to calculate the weight of K in the formulation, the weight of K_2O can be multiplied by 0.83 to compensate for the weight of the oxygen in the molecule. For example, a bag of 10-10-10 has 10 pounds of K_2O, but only 8.3 pounds of K. As an example, the fertilizer *potash* is composed of 52 per cent potassium and 48 per cent chlorine by weight; chemical analysis of 100g of potassium chloride, would show 63g of equivalent potassium oxide when done in the manner of fertilizer analysis. The *percentage yield* of K_2O from the original 100g of fertilizer is the number shown on the label. A potash fertilizer would thus be labelled 0-0-63, and *0-0-52.*

Forms

Fertilizers come in various forms. The most typical form is solid fertilizer in granulated or powdered. The next most common form is liquid fertilizer; some advantages of liquid fertilizer are its immediate effect and wide coverage. There are also slow-release fertilizers which reduce the problem of "burning" the plants due to excess nitrogen. Polymer coating of fertilizer ingredients gives tablets and spikes a "true time-release" or *"staged nutrient release"* (SNR) of fertilizer nutrients.

More recently, organic fertilizer is on the rise as people are resorting to environmental friendly products. Although organic fertilizer usually contain less nutrients, some people still prefer organic due to natural ingredients.

INORGANIC COMMERCIAL FERTILIZER

Fertilizers are broadly divided into organic fertilizers or inorganic or commercial fertilizers. Plants can only absorb their required nutrients if they are present in easily dissolved chemical compounds. Both organic and inorganic fertilizers provide the same needed chemical compounds. Organic fertilizers provided other macro and micro plant nutrients and are released as the organic matter decays—this may take months or years. Organic fertilizers nearly always have much lower concentrations of plant nutrients and have the usual problems of economical collection, treatment, transportation and distribution.

Inorganic fertilizers nearly always are readily dissolved and unless added have few other macro and micro plant nutrients. Nearly all nitrogen that plants use are in the form of NH_3 or NO_3 compounds. The usable phosphorus compounds are usually in the form of phosphoric acid (H_3PO_4) and the potassium (K) is typically in the form of potassium chloride (KCl). In organic fertilizers nitrogen, phosphorus and potassium compounds are released from the complex organic compounds as the animal or plant decays.

In commercial fertilizers the same required compounds are available in easily dissolved compounds that require no decay—they can be used almost immediately after water is applied. Inorganic fertilizers are usually much more concentrated with up to 64 per cent of their weight being a given plant nutrient, compared to organic fertilizers that typically often have only 0.4 per cent or less of a their weight a given plant nutrient.

Nitrogen fertilizers are often made using the Haber-Bosch process which uses natural gas (CH_{4+}) for the hydrogen and nitrogen gas (N_2) from the air at an elevated temperature and pressure in the presence of a catalyst to form ammonia (NH_3) as the end–product. This ammonia is used as a feedstock for other nitrogen fertilizers, such as anhydrous ammonium nitrate (NH_4NO_3) and urea ($CO(NH_2)_2$). These concentrated products may be diluted with water to form a concentrated liquid fertilizer. Deposits of sodium nitrate ($NaNO_3$) are also found the Atacama desert in Chile and was one of the original nitrogen rich inorganic fertilizers used.

It is still mined for fertilizer. In the Nitrophosphate process or Odda Process, phosphate rock with up to a 20 per cent phosphorus (P) content is dissolved with nitric acid (HNO_3) to produce a mixture of phosphoric acid (H_3PO_4) and calcium nitrate ($Ca(NO_3)_2$). This can be combined with a potassium fertilizer to produce a *compound fertilizer* with all three N:P:K: plant nutrients in easily dissolved form.

Phosphate rock can also be processed into water-soluble phosphate (P_2O_5) with the addition of sulfuric acid (H_2SO_4) to make the phosphoric acid in phosphate fertilizers. Phosphate can also be reduced in an electric furnace to make high purity phosphorous; however, this is more expensive than the acid process. Potash can be used to make potassium (K) fertilizers.

All commercial potash deposits come originally from marine deposits and are often buried deep in the earth.

Potash ores are typically rich in potassium chloride (KCl) and sodium chloride (NaCl) and are obtained by conventional shaft mining with the extracted ore ground into a powder. For deep potash deposits hot water is injected into the potash which is dissolved and then pumped to the surface where it is concentrated by solar induced evaporation. Amine reagents are then added to either the mined or evaporated solutions. The amine coats the KCl but not NaCl.

Air bubbles cling to the amine + KCl and float it to the surface while the NaCl and clay sink to the bottom. The surface is skimmed for the amine + KCl which is then dried and packaged for use as a K rich fertilizer—KCl dissolves readily in water and is available quickly for plant nutrition. Compound fertilizers often combine N, P and K fertilizers into easily dissolved pellets. The N:P:K ratios quoted on fertilizers give the weight per cent of the fertilizer in nitrogen (N), phosphate (P_2O_5) and potash (K_2O equivalent).

The use of commercial inorganic fertilizers has increased steadily in the last 50 years, rising almost 20-fold to the current rate of 100 million tonnes of nitrogen per year. Without commercial fertilizers it is estimated that about one-third of the food produced now could not be produced. The use of phosphate fertilizers has also increased from 9 million tones per year in 1960 to 40 million tones per year in 2000. A maize crop yielding 6–9 tones of grain per hectare requires 31–50 kg of phosphate fertilizer to be applied, soybean requires 20–25 kg per hectare. Yara International is the world's largest producer of nitrogen based fertilizers.

Controlled-release Types

Urea and formaldehyde, reacted together to produce sparingly soluble polymers of various molecular weights, is one of the oldest controlled-nitrogen-release technologies, having been first produced in 1936 and commercialized in 1955. The early product had 60 per cent of the total nitrogen cold-water-insoluble, and the unreacted less than 15 per cent. Methylene ureas were commercialized in the 1960s and 1970s, having 25 and 60 per cent of the nitrogen cold-water-insoluble, and unreacted urea nitrogen in the range of 15 to 30 per cent.

Isobutylidene diurea, unlike the methylurea polymers, is a single crystalline solid of relatively uniform properties, with about 90 per cent of the nitrogen water-insoluble. In the 1960s the National Fertilizer Development Centre began developing Sulfur-coated urea; sulfur was used as the principle coating material because of its low cost and its value as a secondary nutrient.

Usually there is another wax or polymer which seals the sulfur; the slow release properties depend on the degradation of the secondary sealant by soil microbes as well as mechanical imperfections in the sulfur. They typically provide 6 to 16 weeks of delayed release in turf applications. When a hard polymer is used as the secondary coating, the properties are a cross between diffusion-controlled particles and traditional sulfur-coated. Other coated products use thermoplastics to produce diffusion-controlled release of urea or soluble inorganic fertilizers.

"Reactive Layer Coating" can produce thinner, hence cheaper, membrane coatings by applying reactive monomers simultaneously to the soluble particles. "Multicote" is a process applying layers of low-cost fatty acid salts with a paraffin topcoat. Besides being more efficient in the utilization of the applied nutrients, slow-release technologies also reduce the impact on the environment and the contamination of the subsurface water.

Application

Synthetic fertilizers are commonly used for growing all crops, with application rates depending on the soil fertility, usually as measured by a soil test and according to the particular crop. Legumes, for example, fix nitrogen from the atmosphere and generally do not require nitrogen fertilizer. Studies have shown that application of nitrogen fertilizer on off-season cover crops can increase the biomass of these crops, while having a beneficial effect on soil nitrogen levels for the main crop planted during the summer season.

Nutrients in soil can be thrown out of balance with high concentrations of fertilizers. The interconnectedness and complexity of this soil "food web" means any appraisal of soil function must necessarily take into account interactions with the living communities that exist within the soil. Stability of the system is reduced by the use of nitrogen-containing fertilizers, which causesoil acidification. Applying excessive amounts of fertilizer has negative environmental effects, and wastes the growers' time and money.

To avoid over-application, the nutrient status of crops should be assessed. Nutrient deficiency can be detected by visually assessing the physical symptoms of the crop. Nitrogen deficiency, for example has a distinctive presentation in some species. However, quantitative tests are more reliable for detecting nutrient deficiency before it has significantly affected the crop. Both soil tests and plant tissue tests are used in agriculture to fine-tune nutrient management to the crops needs.

Problems with Inorganic Fertilizer

Common agricultural grade phosphate fertilizers usually contain impurities such as fluorides, cadmium and uranium, although concentrations of the latter two heavy metals are dependent on the source of the phosphate and the production process. These potentially harmful impurities can be removed; however, this significantly increases cost.

Highly pure fertilizers are widely available and perhaps best known as the highly water soluble fertilizers containing blue dyes used around households. These highly water soluble fertilizers are used in the plant nursery business and are available in larger packages at significantly less cost than retail quantities. There are also some inexpensive retail granular garden fertilizers made with high purity ingredients.

Oregon and Washing in US have fertilizer registration programmes with on-line databases listing chemical analyses of fertilizers. The most widely used inorganic fertilizer is super-phosphate and its double and triple strengthed derivatives double super and triple super. Superphosphate was first developed by Lawes at the Rothamstead Agricultural Research Institute in England in the early 19th century. Lawes added sulfuric acid to conventional rock phosphate containing the mineral apatite, a calcium fluoro-phosphate.

The resulting water soluble phosphorus was able to significantly improve yields on a variety of crops at the Rothamstead Centre and the Superphosphate industry was born. Unfortunately over decades of subsequent useage-it became clear that the solubilisation of fluorine also occurred in the process and this had the same effect as the other halogen sterilants over time-soil sterilization.

Effectively farmers unknowingly became 100 per cent dependent on "bought in" water soluble, inorganic fertilizers since the sterilization of soil microflora including its micorhizza, reduced the availability of other natural and trace minerals within the soil. This to some extent explains the resurgence of interest in organic and particularly "biodynamic" farming systems since these systems replace the essential soil organisms so essential to converting soil minerals into plant available nutrients. They do this by a variety of processes including chelation whereby essential minerals become plant available–as measured by weak citric acid extraction techniques.

Hence the citric acid solubility of phosphate rocks has emerged as a measure of plant availability and enabled so-called "reactive" phosphate rocks to be used as fertilizer minerals. These should not be confused with high fluorine apatite rocks in which the fluoride content performs a similar function to its role in hardening teeth enamel, *i.e.,* immobilizing phosphorus. This explains the oceanic origins of many of these high fluorine rocks since the fluorine absorbed from the sea has prevented what were originally massive deposits of bird guano-from being leached from the coral based limestone rocks on which they were originally deposited. Also regular use of acidulated fertilizers generally contribute to the accumulation of soil acidity in soils which progressively increases aluminium availability and hence toxicity. The use of such acidulated fertilizers in the tropical and semi-tropical regions of Indonesia and Malaysia has contributed to soil degredation on a large scale from aluminium toxicity, which can only be countered by applications of limestone or preferably magnesian dolomite, which neutralises acid soil pH and also provides essential magnesium.

Trace Mineral Depletion

Many inorganic fertilizers, particularly those based on superphosphate, may not replace trace mineral elements in the soil which become gradually depleted by crops. This depletion has been linked to studies which have shown a marked fall in the quantities of such minerals present in fruit and vegetables.Explanations for this include the early encouragement of so-called "luxury consumption" of trace elements as a result of their acidulation and subsequent dissolution in soil water, by free sulphuric acid sourced from superphosphate. This mechanism has also been identified as a possible causal agent for take-up of the heavy metal cadmium from superphosphate based fertilizers. In Western Australia deficiencies of zinc, copper, manganese, iron and molybdenum were identified as limiting the growth of broad-acre crops and pastures in the 1940s and 1950s. Such nutrients are described as "rate limiting" nutrients.

Soils in Western Australia are very old, highly weathered and deficient in many of the major nutrients and trace elements. Since this time these trace elements are routinely added to inorganic fertilizers used in agriculture in this state. Many soils around the world are deficient in zinc, leading to deficiency in plants and humans.

Over-fertilization

Over-fertilization of a vital nutrient can be as detrimental as under–fertilization. "Fertilizer burn" can occur when too much fertilizer is applied, resulting in a drying out of the leaves and damage or even death of the plant. Fertilizers vary in their tendency to burn roughly in accordance with their salt index.

High Energy Consumption

In the USA in 2004, 317 billion cubic feet of natural gas was consumed in the industrial production of ammonia, less than 1.5 per cent of total US annual consumption of natural gas. A 2002 report suggested that the production of ammonia consumes about 5 per cent of global natural gas consumption, which is somewhat under 2 per cent of world energy production. Ammonia is overwhelmingly produced from natural gas, but other energy sources, together with a hydrogen source, can be used for the production of nitrogen compounds suitable for fertilizers. The cost of natural gas makes up about 90 per cent of the cost of producing ammonia. The increase in price of natural gases over the past decade, along with other factors such as increasing demand, have contributed to an increase in fertilizer price.

Long-term Sustainability

Inorganic fertilizers are now produced in ways which theoretically cannot be continued indefinitely by definition as the resources used in their production are non-renewable. Potassium and phosphorus come from mines and such resources are limited. However, more effective fertilizer utilization practices may decrease present usage from mines.

Improved knowledge of crop production practices can potentially decrease fertilizer usage of P and K without reducing the critical need to improve and increase crop yields. Atmospheric nitrogen is effectively unlimited but this is not in a form useful to plants. To make nitrogen accessible to plants requires nitrogen fixation.

Artificial nitrogen fertilizers are typically synthesized using fossil fuels such as natural gas and coal, which are limited resources. In lieu of converting natural gas to syngas for use in the Haber process, it is also possible to convert renewable biomass to syngas to supply the necessary energy for the process, though the amount of land and resources necessary for such a project may be prohibitive.

FARMING MANAGEMENT PRACTICES

Farming management practices include land preparation, planting, thinning and weeding, fertilizer application, pest control, irrigation, harvesting, and post-harvest processing. Each of the processes mayor may not affect the amount of genetic diversity of the crop population over time. Different levels of fertilizer application or the use of organic rather than chemical fertilizer may select for different genotypes in a population. Irrigated and non-irrigated populations of faba beans in Morocco have different population genetic characteristics. Dense planting to reduce weeding and post-harvest storage of seeds also play a selection role on seed survival and the continuation of subsequent traits in the next generation.

Key questions for national on-farm conservation programmes are (1) which farming management practices influence genetic diversity and (2) to what extent do these practices affect the amount of genetic variation in the crop population. For national programmes to look at these questions, interviews and observations of farmer management practices are needed. Field trials using different degrees of density or fertilizer application coupled with measurements of genetic diversity may be desirable to investigate specific effects of particular treatments. Such studies may, however, lie outside the capacity of this particular project and be best explored in specific case studies.

Environmental selection: When a farmer decides to plant a cultivar on a particular micro-site, the crop population is exposed to specific environmental selection. By planting barley in soils prone to water logging or in different agroecological sites, the Ethiopian farmer is subjecting the plants to environmental selection for tolerance to a specific stress. Similarly, by planting upland "Jumla" rice at elevations up to 3000 m in Nepal, or "Chao" rice on shallow soils in northern Vietnam, the farmer is exposing a crop population to cold stress or poor soil conditions.

Questions under this selection category revolve around (1) how environmental selection has influenced the genetic diversity of the population over time and (2) which environmental conditions playa significant role in affecting the amount of genetic variation in the crop population. The farmer's involvement is in making the decision to plant that particular population in a particular micro-habitat.

To answer these questions, national programmes will need to focus on basic questions of ecological genetics, to understand how populations adapt to their environments, and to determine the type and extent of effect of environmental factors on the amount of genetic variation over time. The extent to which farmers maintain unique types for particular micro-environments will be important, as well as the occurrence within local cultivars of G × E interactions or broad tolerance of a range of stress environments. Social scientists' involvement will be in investigating what factors influence a farmer to plant in the particular habitat. The work will likely point to specific studies over longer periods that will help breeders and users understand the nature of adaptation in local cultivars. *Population size:* The size of a population planted by a farmer will affect the amount of genetic variation of the crop population over time. The smaller the population, the more likely it is that genetic drift, inbreeding, loss of alleles, and stochastic events will affect the population. From a conservation perspective, crucial questions include: How does population size influence the genetic diversity of the local cultivar population? What is the effective population size that ensures long-term stability of the population? Should a national programme be concerned with a group of smaller populations or metapopulations, or should the conservation focus be on individual farmer fields? To understand the effect of size on the amount of genetic variation of a population requires minimum population viability studies and population genetic methods.

Seed source

A farmer makes a choice each year on what percentage of his or her own seeds to save and plant and what percentage he or she will acquire from other farmers. How do migration (influx of new seeds) and bottlenecks (reduction in the number of saved seeds) affect the genetic diversity of the population over time? Louette and Smale (1996) have shown that after six crop rotations of the farmer studies in Mexico, only 48 per cent of the seed material remained from the farmer's original stock. This is the information that a population geneticists will need to determine effective population sizes. Again, the question of whether a metapopulation should be the unit of concern for national conservation programmes is important. This involves social scientists and community-based groups developing an understanding of the seed supply system in order for plant geneticists to be able to determine the effective population size for the cultivar in question. This requires knowledge, not only of seed source and informal seed supply systems, but also of how different storage systems influence the survival of different seeds over time.

On-farm data for conservation and use: On-farm maintenance of local cultivars is currently being promoted as a strategy to conserve genetic diversity and as a process that will secure the continued availability of genetic variation while ensuring the continuing betterment of farmer livelihood over time. In the past, most conservation workers have focused largely on *ex situ* conservation, with the expectation that local cultivars maintained over the centuries will shortly disappear. This is now seen as an oversimplification of the complex issues at play (Louette, this volume). Knowledge of the processes involved in farmer selection and maintenance of crop cultivars can be used by national programmes to make decisions on where to support, assist, or intervene to promote the conservation of genetic diversity and strengthen the effectiveness of conservation. How can this same information, linking farmer decision making to genetic diversity, be used to help national programmes improve the livelihood of the farmer? To be maintained by farmers, crop genetic resources must have value to them and must retain this value and be competitive to other options a farmer might have, or even better, to increase this value such that it is reflected in an increase in a farmer's standard of living.

Value may be added to crop resources in two main ways: the material itself may be improved or the demand for the material or some product may increase. One option is to seek improved quality, disease resistance, yield, taste, or other farmer preferred characteristics, through decentralized breeding activities. Sthapit, Joshi, and Witcombe (1996a) have shown that, by utilizing farmers' knowledge, acceptable varieties can be bred with minimum use of resources in the high altitude areas of Nepal. Effective support for farm based maintenance needs to explore the ways in which this can best be done and the impact it has on the extent and distribution of genetic diversity.

Value can also be added to crop resources by better processing, storage, and marketing, where the farmer receives more benefit from the final product. An important role for community-based organizations and nongovernmental organizations is the formation of farmer cooperatives and farmer-managed community seed banks to maximize returns to the farmers themselve. Government policy may also playa role in ensuring that farmers' inputs and seeds receive the same or better market treatment and government support as improved varieties. *Conclusion:* The majority of early studies on on-farm conservation have focused on trying to determine what factors have caused a farmer to maintain or not maintain a diversity of local cultivars. Other studies have looked at the amount of genetic variation of these cultivars at a given moment in time. The research discussed in this chapter aims to take earlier studies further by attempting to quantify over time how specific farmer-based decisions may determine both the amount of genetic variation and the effective population size of a cultivar population over time.

Changes to the amount of genetic variation can be measured as can the effect of farmer-based management on the effective population sizes. Genetic variation may be linked to farmer decisions in five major categories: agromorphological characteristics, farm management practices, planting location, size of the population, and seed source. These decisions in turn are based on environmental and socioeconomic influences. The first link forms the basic data set for the second link. Understanding these processes requires the involvement of people from informal and formal institutes and from a variety of scientific disciplines. Once there is better knowledge of these processes, national programmes will be in a better position to support, assist, or intervene in the conservation of genetic diversity on-farm.

Farmers themselves will ultimately determine the extent to which on-farm maintenance of local cultivars continues and, hence, contributes to the overall conservation of crop genetic diversity. Previous conservation work underplayed or even ignored this contribution and emphasized *ex situ* methods of diversity maintenance. Through the creation of a framework of knowledge of the processes involved, it is hoped to provide a more complete understanding that redresses this imbalance.

INTEGRATED FARMING SYSTEMS

Development and adoption of integrated farming system provides high opportunities of productivity enhancement, employment, income generation and nutritional security by diversifying and integrating different components of farming, *viz.* crops, horticulture, livestock and fisheries (depending upon location specificity). The systems based on multiple recycling of carbon, energy and nutrients would also help minimize environmental loading with pollutants. The different components of the system have complementarities with waste products of one component becoming source of food and energy for another. The researchable issues encompassing analysis of nutrient and energy fluxes among system components, path analysis of bio-physical constraints, multiple uses of water, water harvesting and recycling, socio-economic conditions of people, internalization of ITK and market analysis, etc. are to be addressed through basic and strategic research by ICAR institutes and SAUs. The integrated technology would be developed integrating crops, horticulture, fisheries & livestock by ICAR Research Complexes having R&D facilities on different system components at SAUs.

Further assessment and refinement of developed farming systems to cater to region and location specific requirements would be accomplished through centres of AICRPs on Cropping/Farming System Research and SAUs. The technology would be disseminated through KVKs, SAUs, NGOs, SHGs, etc. with true participation of ATMAs and different functionaries of State Line Departments (Agriculture, Horticulture, Animal Husbandry, Fishery) at district and block levels. The backward and forward linkages between KVKs and state functionaries would provide the required feedback to SAUs and ICAR institutes on upgradation of technology. The KVKs would organize regular trainings on different aspects of technology to extension personnel and various stakeholders.

ESTABLISHMENT OF AGRI-INDIA KNOWLEDGE PORTAL: A SINGLE ELECTRONIC GATEWAY

The information on improving agricultural productivity and protection of crops, livestock and natural resources from damages caused by disasters and unsustainable activities is focal to the rural farming communities. This would require access to a wide range of information on new technologies, alternate varieties of crops, improved breeds of livestock, information related to soil, water quality, information on pesticides, farm implements, animal health, weather and other related aspects. Besides farm-activities, non-farm activities also constitute sources of income and livelihood for the rural farmers. Therefore information is also required on the means to produce food products required by markets and ability to sell them as also the markets, cold-chains, warehouses, processing and other avenue.

ICAR-ERNET network consists of 274 Institutions including National Agriculture Institutes, State Agricultural Universities including some of their colleges and Research stations, National Research Centres, Project Directorates connected through Leased line and VSATs. Besides, a VSAT-based network of Krishi Vigyan Kendras could be had. Both these networks will be linked through a high capacity leased line to make a unified high speed secured Intranet. Since the intranet of NARS institutions is built over the existing ERNET network, it is planned to host the core components of knowledge portal application at the ERNET Hub. The Department of Telecommunications has set up several sub-groups on network expansion, broadband, telecom equipment manufacturing and R&D to examine various options to enhance growth. The sub group on network expansion has targeted that subscribers numbers must cross 250 million by 2007 and 500 million by 2010. Another target is mobile coverage of 85 per cent of the country by the end of 2007. To boost the broadband penetration in the country, another sub-group will examine the ways to provide broadband coverage for all panchayats by 2010 and all secondary/ higher secondary schools and public health care centres by 2007. The Portal application will be constructed and deployed using suitable Portal management software. The core functional components of the portal application software are:

User Registration: The user registration process is a mechanism for acquiring the information related with the user of portal which is an integral part of the portal.

User Authentication: The user ID and the password pair generated as end result of registration process are authenticated through a carefully designed mechanism and form a basis for accessing the various resources of a portal.

User Access Control: Once the users are authenticated, an access control mechanism built into the portal determines which data and information can they see/manipulate, and how will it be displayed? The User Access Control mechanism is designed based upon the specific role that the particular user will perform in the portal.

Personalization: Personalization is a feature that enables a user to access targeted web content that meets a user's needs and preferences from the enormous web content on the portal.

Content and Document Management: Content and document management system in the portal application streamline redundant content development, publishing and management processes to increase operational efficiencies, and improve content quality.

Document/workflow: The various rules to manage the workflow and lifecycle of documents are established in a portal, which automates the manual, labor-intensive process and results in consistency and auditability in authoring and producing content.

Collaboration: The portal applications facilitate collaboration among the various stakeholders/ partners via online chat rooms, discussion groups, online document sharing, messenger applications, etc. The digital content of the portal will be developed through a peer review process. For this purpose, 15 Content Accreditation Centres, *i.e.* one each in the 15 Agro-climatic regions of the country could be had. Each accreditation centre will coordinate with other SAUs and agricultural institutions in their region for development of content in regional language as well as in English and also do its validation, which will be collected in the central data warehouse integrated in the knowledge portal.

The intranet network consisting of various stakeholders primarily ICAR institutes, SAUs and KVKs will facilitate the development of Central Agriculture Data warehouse, Content Development and Validation System, Agricultural Best Practices, Expert System for users, Multi-lingual content development and convertibility. The portal will also serve as a platform for facilitation of interaction among researchers and extension workers in KVKs through high speed secure intranet.

Various applications and services of the knowledge portal will be accessible to the users including farmers, entrepreneurs, private sector organizations, other participating government and non-government organizations, extension workers, etc. through Internet. The services include Agro-met advisory services, Market intelligence, Packages of practices, Agri-business consultancy, e-learning, knowledge on indigenous farming practices, and agriculture related FAQs. Besides, it will also act as information gateway for all agriculture related schemes and programmes etc. The Portal will also serve as a gateway for up-to-date websites of various agricultural research and education organizations accessed through Internet. The portal will provide a platform for facilitation of interaction among the end user communities through Internet also.

The Agri-India Knowledge portal will also provide for collaboration among other agencies related with agriculture like Meteorological Dept, Agricultural Markets, Banks and other financial organization, Input suppliers, Agricultural equipment manufacturers, Non-agricultural universities for providing services through Agri-India knowledge portal both through secure and high speed Intranet and Internet. The services proposed to be implemented in the collaborated mode include On-line education, Information on Market Prices, Weather Information including early warning system, Agricultural Input status information, Credit and crop insurance services, land record information, etc.

The AIK Portal is neither intended to replace any agency websites nor any other government portal. This Portal will, however, provide additional means by which those sites might be accessed by a possibly wider and more varied client base. The AIK Portal will open opportunities for multiple agencies to participate in web delivery and development opportunities. The Portal would be able to integrate diverse interaction channels at a central point, providing a comprehensive context and an aggregated views across all information related to agriculture including *know-what* or declarative knowledge, *know-how* or procedural knowledge, and *know-why* or usual knowledge and *creation of new knowledge.*

FARMING TECHNIQUES IN SUSTAINABLE AGRICULTURE

Sustainable agriculture is a special kind of agriculture along with farming technique which makes maximum utilization of the environment without causing any form of harm to it. The outputs from this Sustainable Agriculture are devoid of any types of inorganic chemicals such as pesticides and insecticides. This farming technique uses the organic way of farming.

All these factors produce the products in a more environment friendly manner and are thus healthier for the consumers to consume.

- Sustainable Agriculture Practice is regarded as the need of the hour by most of the countries over the world keeping in view environmental concern.
- Sustainable Agriculture Washington is an organization, which has been rendering promotional methods in the field of agriculture. Find more about Sustainable Agriculture in Washington.
- National Sustainable Agriculture is one of the leading national level programme in United States of America. Get more on National Sustainable Agriculture.

The other characteristic feature of Sustainable Agriculture are given in the bullet form below:

- The treatment towards the animals in such kind of farming are more humane because they are given enough space to graze and are not bred like factory products.
- The workers who are associated with the Sustainable Agriculture receive more humane behavior from the employers and it has been observed that the wages offered to these farm workers are fairer than the ones working in the so called industrial farms.
- The one of the most important feature of Sustainable Agriculture is its extended support towards the small farming community farmers. It is essentially being emphasized in this type of farming that the organic food items that have been produced by the small farmers

of the rural areas are bought and marketed to the consumers through this Sustainable Agriculture route.

Sustainable Agriculture got its first legal recognition in the United States of America through the Public Law 101-624, Title XVI, Subtitle A, Section 1683 which was introduced in the year 1990. This law defined this farming practice type to be more Eco-friendly and having a sight for the long run satisfaction of the nutrient values in accordance with the nutritional requirements of the human body. It was especially emphasized in the law that the non renewable resources that are used in the farming process needs to be used at a minimal level in Sustainable Agriculture. It emphasizes that the viability conditions of the small farmers in the rural areas needs to be enhanced through Sustainable Agriculture through proper disposal of wage and proper distribution of their produced products.

Information on Sustainable Agriculture would be incomplete if the mention of the Organic Farming is not considered in the same. For more development in this field, research on sustainable agriculture is going on and with every passing day, this farming type is given more emphasis by the American as well as the French government.

FARMING AND NATURAL RESOURCES

Water. When the production of food and fiber degrades the natural resource base, the ability of future generations to produce and flourish decreases. The decline of ancient civilizations in Mesopotamia, the Mediterranean region, Pre-Columbian southwest U.S. and Central America is believed to have been strongly influenced by natural resource degradation from non-sustainable farming and forestry practices. Water is the principal resource that has helped agriculture and society to prosper, and it has been a major limiting factor when mismanaged.

Water supply and use. In California, an extensive water storage and transfer system has been established which has allowed crop production to expand to very arid regions. In drought years, limited surface water supplies have prompted overdraft of groundwater and consequent intrusion of salt water, or permanent collapse of aquifers. Periodic droughts, some lasting up to 50 years, have occurred in California.

Several steps should be taken to develop drought-resistant farming systems even in "normal" years, including both policy and management actions:

- Improving water conservation and storage measures,
- Providing incentives for selection of drought-tolerant crop species,
- Using reduced-volume irrigation systems,
- Managing crops to reduce water loss, or
- Not planting at all.

Water quality: The most important issues related to water quality involve salinization and contamination of ground and surface waters by pesticides, nitrates and selenium. Salinity has become a problem wherever water of even relatively low salt content is used on shallow soils in arid regions and/or where the water table is near the root zone of crops. Tile drainage can remove the water and salts, but the disposal of the salts and other contaminants may negatively affect the environment depending upon where they are deposited.

Temporary solutions include the use of salt-tolerant crops, low-volume irrigation, and various management techniques to minimize the effects of salts on crops. In the long-term, some farmland may need to be removed from production or converted to other uses. Other uses include conversion of row crop land to production of drought-tolerant forages, the restoration of wildlife habitat or the use of agroforestry to minimize the impacts of salinity and high water tables. Pesticide and nitrate contamination of water can be reduced using many of the practices discussed later in the Plant Production Practices.

Wildlife: Another way in which agriculture affects water resources is through the destruction of riparian habitats within watersheds. The conversion of wild habitat to agricultural land reduces fish and wildlife through erosion and sedimentation, the effects of pesticides, removal of riparian plants, and the diversion of water. The plant diversity in and around both riparian and agricultural areas should be maintained in order to support a diversity of wildlife. This diversity will enhance natural ecosystems and could aid in agricultural pest management.

Energy: Modern agriculture is heavily dependent on non-renewable energy sources, especially petroleum. The continued use of these energy sources cannot be sustained indefinitely, yet to abruptly abandon our reliance on them would be economically catastrophic. However, a sudden cutoff in energy supply would be equally disruptive. In sustainable agricultural systems, there is reduced reliance on non-renewable energy sources and a substitution of renewable sources or labour to the extent that is economically feasible.

Air: Many agricultural activities affect air quality. These include smoke from agricultural burning; dust from tillage, traffic and harvest; pesticide drift from spraying; and nitrous oxide emissions from the use of nitrogen fertilizer. Options to improve air quality include incorporating crop residue into the soil, using appropriate levels of tillage, and planting wind breaks, cover crops or strips of native perennial grasses to reduce dust.

Soil: Soil erosion continues to be a serious threat to our continued ability to produce adequate food. Numerous practices have been developed to keep soil in place, which include reducing or eliminating tillage, managing irrigation to reduce runoff.

CONCEPT THEMES

Sustainable agriculture integrates three main goals—environmental health, economic profitability, and social and economic equity. A variety of philosophies, policies and practices have contributed to these goals. People in many different capacities, from farmers to consumers, have shared this vision and contributed to it. Despite the diversity of people and perspectives, the following themes commonly weave through definitions of sustainable agriculture.

Sustainability rests on the principle that we must meet the needs of the present without compromising the ability of future generations to meet their own needs. Therefore, *stewardship of both natural and human resources* is of prime importance. Stewardship of human resources includes consideration of social responsibilities such as working and living conditions of laborers, the needs of rural communities, and consumer health and safety both in the present and the future. Stewardship of land and natural resources involves maintaining or enhancing this vital resource base for the long term.

A *systems perspective* is essential to understanding sustainability. The system is envisioned in its broadest sense, from the individual farm, to the local ecosystem, *and* to communities affected by this farming system both locally and globally. An emphasis on the system allows a larger and more thorough view of the consequences of farming practices on both human communities and the environment. A systems approach gives us the tools to explore the interconnections between farming and other aspects of our environment.

A systems approach also implies *interdisciplinary efforts in research and education*. This requires not only the input of researchers from various disciplines, but also farmers, farmworkers, consumers, policymakers and others.

Making the transition to sustainable agriculture is a process. For farmers, the transition to sustainable agriculture normally requires a series of small, realistic steps. Family economics and personal goals influence how fast or how far participants can go in the transition. It is important to realize that each small decision can make a difference and contribute to advancing the entire system further on the "sustainable agriculture continuum." The key to moving forward is the will to take the next step.

Finally, it is important to point out that *reaching toward the goal of sustainable agriculture is the responsibility of all participants in the system,* including farmers, laborers, policymakers, researchers, retailers, and consumers. Each group has its own part to play, its own unique contribution to make to strengthen the sustainable agriculture community.

The remainder of this document considers specific strategies for realizing these broad themes or goals. The strategies are grouped according to three separate though related areas of concern: Farming and Natural Resources, Plant and Animal Production Practices, and the Economic, Social and Political Context. They represent a range of potential ideas for individuals committed to interpreting the vision of sustainable agriculture within their own circumstances.

PLANT PRODUCTION PRACTICES

Sustainable production practices involve a variety of approaches. Specific strategies must take into account topography, soil characteristics, climate, pests, local availability of inputs and the individual grower's goals.

Despite the site-specific and individual nature of sustainable agriculture, several general principles can be applied to help growers select appropriate management practices:

- Selection of species and varieties that are well suited to the site and to conditions on the farm;
- Diversification of crops and cultural practices to enhance the biological and economic stability of the farm;
- Management of the soil to enhance and protect soil quality;
- Efficient and humane use of inputs; and
- Consideration of farmers' goals and lifestyle choices.

Selection of site, species and variety: Preventive strategies, adopted early, can reduce inputs and help establish a sustainable production system. When possible, pest-resistant crops should be selected which are tolerant of existing soil or site conditions. When site selection is an option, factors such as soil type and depth and location should be taken into account before planting.

Diversity: Diversified farms are usually more economically and ecologically resilient. While monoculture farming has advantages in terms of efficiency and ease of management, the loss of the crop in any one year could put a farm out of business and/or seriously disrupt the stability of a community dependent on that crop. By growing a variety of crops, farmers spread economic risk and are less susceptible to the radical price fluctuations associated with changes in supply and demand.

Properly managed, diversity can also buffer a farm in a biological sense. In annual cropping systems, crop rotation can be used to suppress weeds, pathogens and insect pests. Also, cover crops can have stabilizing effects on the agroecosystem by holding soil and nutrients in place, conserving soil moisture with mowed or standing dead mulches, and by increasing the water infiltration rate and soil water holding capacity. Cover crops in orchards and vineyards can buffer the system against pest infestations by increasing beneficial arthropod populations and can therefore reduce the need for chemical inputs. Using a variety of cover crops is also important in order to protect against the failure of a particular species to grow and to attract and sustain a wide range of beneficial arthropods. Optimum diversity may be obtained by integrating both crops and livestock in the same farming operation. This was the common practice for centuries until the mid-1900s when technology, government policy and economics compelled farms to become more specialized. Mixed crop and livestock operations have several advantages. First, growing row crops only on more level land and pasture or forages on steeper slopes will reduce soil erosion. Second, pasture and forage crops in rotation enhance soil quality and reduce erosion; livestock manure, in turn, contributes to soil fertility. Third, livestock can buffer the negative impacts of low rainfall periods by consuming crop residue that in "plant only" systems would have been considered crop failures. Finally, feeding and marketing are flexible in animal production systems. This can help cushion farmers against trade and price fluctuations and, in conjunction with cropping operations, make more efficient use of farm labour.

Soil management: A common philosophy among sustainable agriculture practitioners is that a "healthy" soil is a key component of sustainability; that is, a healthy soil will produce healthy crop plants that have optimum vigor and are less susceptible to pests. While many crops have key pests that attack even the healthiest of plants, proper soil, water and nutrient management can help prevent some pest problems brought on by crop stress or nutrient imbalance. Furthermore, crop management systems that impair soil quality often result in greater inputs of water, nutrients, pesticides, and/or energy for tillage to maintain yields.

In sustainable systems, the soil is viewed as a fragile and living medium that must be protected and nurtured to ensure its long-term productivity and stability. Methods to protect and enhance the productivity of the soil include using cover crops, compost and/or manures, reducing tillage, avoiding traffic on wet soils, and maintaining soil cover with plants and/or mulches. Conditions in most California soils do not favor the buildup of organic matter. Regular additions of organic matter or the use of cover crops can increase soil aggregate stability, soil tilth, and diversity of soil microbial life.

Efficient use of inputs: Many inputs and practices used by conventional farmers are also used in sustainable agriculture. Sustainable farmers, however, maximize reliance on natural, renewable, and on-farm inputs. Equally important are the environmental, social, and economic impacts of a particular strategy. Converting to sustainable practices does not mean simple input substitution. Frequently, it substitutes enhanced management and scientific knowledge for conventional inputs, especially chemical inputs that harm the environment on farms and in rural communities. The goal is to develop efficient, biological systems which do not need high levels of material inputs.

Growers frequently ask if synthetic chemicals are appropriate in a sustainable farming system. Sustainable approaches are those that are the least toxic and least energy intensive, and yet maintain productivity and profitability. Preventive strategies and other alternatives should be employed before using chemical inputs from any source. However, there may be situations where the use of synthetic chemicals would be more "sustainable" than a strictly non-chemical approach or an approach using toxic "organic" chemicals. One grape grower switched from tillage to a few applications of a broad spectrum contact herbicide in the vine row. This approach may use less energy and may compact the soil less than numerous passes with a cultivator or mower.

Consideration of farmer goals and lifestyle choices. Management decisions should reflect not only environmental and broad social considerations, but also individual goals and lifestyle choices. Adoption of some technologies or practices that promise profitability may also require such intensive management that one's lifestyle actually deteriorates. Management decisions that promote sustainability, nourish the environment, the community *and* the individual.

ANIMAL PRODUCTION PRACTICES

In the early part of this century, most farms integrated both crop and livestock operations. Indeed, the two were highly complementary both biologically and economically. The current picture has changed quite drastically since then. Crop and animal producers now are still dependent on one another to some degree, but the integration now most commonly takes place at a higher level—*between* farmers, through intermediaries, rather than *within* the farm itself. This is the result of a trend toward separation and specialization of crop and animal production systems. Despite this trend, there are still many farmers, particularly in the Midwest and Northeastern U.S. that integrate crop and animal systems—either on dairy farms, or with range cattle, sheep or hog operations. Even with the growing specialization of livestock and crop producers, many of the principles outlined in the crop production section apply to both groups. The actual management practices will, of course, be quite different. Some of the specific points that livestock producers need to address are listed below.

Management Planning: Including livestock in the farming system increases the complexity of biological and economic relationships. The mobility of the stock, daily feeding, health concerns, breeding operations, seasonal feed and forage sources, and complex marketing are sources of this complexity. Therefore, a successful ranch plan should include enterprise calendars of operations, stock flows, forage flows, labour needs, herd production records and land use plans to give the manager control and a means of monitoring progress toward goals.

Animal Selection: The animal enterprise must be appropriate for the farm or ranch resources. Farm capabilities and constraints such as feed and forage sources, landscape, climate and skill of the manager must be considered in selecting which animals to produce. Ruminant animals can be raised on a variety of feed sources including range and pasture, cultivated forage, cover crops, shrubs, weeds, and crop residues. There is a wide range of breeds available in each of the major ruminant species, *i.e.*, cattle, sheep and goats. Hardier breeds that, in general, have lower growth and milk production potential, are better adapted to less favorable environments with sparse or highly seasonal forage growth.

Animal nutrition: Feed costs are the largest single variable cost in any livestock operation. While most of the feed may come from other enterprises on the ranch, some purchased feed is usually imported from off the farm. Feed costs can be kept to a minimum by monitoring animal condition and performance and understanding seasonal variations in feed and forage quality on the farm. Determining the optimal use of farm-generated by-products is an important challenge of diversified farming.

Reproduction: Use of quality germplasm to improve herd performance is another key to sustainability. In combination with good genetic stock, adapting the reproduction season to fit the climate and sources of feed and forage reduce health problems and feed costs.

Herd Health: Animal health greatly influences reproductive success and weight gains, two key aspects of successful livestock production. Unhealthy stock waste feed and require additional labour. A herd health programme is critical to sustainable livestock production.

Grazing Management: Most adverse environmental impacts associated with grazing can be prevented or mitigated with proper grazing management. First, the number of stock per unit area must be correct for the landscape and the forage sources. There will need to be compromises between the convenience of tilling large, unfenced fields and the fencing needs of livestock operations.. Second, the long term carrying capacity and the stocking rate must take into account short and long-term droughts. Especially in Mediterranean climates such as in California, properly managed grazing significantly reduces fire hazards by reducing fuel build-up in grasslands and brushlands. Finally, the manager must achieve sufficient control to reduce overuse in some areas while other areas go unused. Prolonged concentration of stock that results in permanent loss of vegetative cover on uplands or in riparian zones should be avoided. However, small scale loss of vegetative cover around water or feed troughs may be tolerated if surrounding vegetative cover is adequate.

Confined Livestock Production: Animal health and waste management are key issues in confined livestock operations. The moral and ethical debate taking place today regarding animal welfare is particularly intense for confined livestock production systems. The issues raised in this debate need to be addressed.

Confinement livestock production is increasingly a source of surface and ground water pollutants, particularly where there are large numbers of animals per unit area. Expensive waste management facilities are now a necessary cost of confined production systems. Waste is a problem of almost all operations and must be managed with respect to both the environment and the quality of life in nearby communities. Livestock production systems that disperse stock in pastures so the wastes are not concentrated and do not overwhelm natural nutrient cycling processes have become a subject of renewed interest.

6

Economic Reforms in Agriculture

Although India's economic reforms were initiated in June 1991, the process of liberalization was implemented gradually and thus it is difficult to assess the full impact of the liberalization measures. Nevertheless, an attempt is made to discuss what is observable in terms of agricultural growth. One observation is that the expected increase in exports due to liberalization simply did not occur. India's share in world exports was 0.6 per cent in 1997; India has to aim for at least 4 per cent by 2005 in order to meet the growing import demands for capital goods, raw materials and crude oil as well as to meet her external financial commitments. For the last decade or so, India's share in world exports of agriculture has been between 2 per cent and 3 per cent. India is not as competitive as the other countries and calculations show that India's crop yields have increased at a slower rate over the 1990s.

In addition, the agricultural sector's output growth decreased to 2.9 per cent during 1992-1993 to 1998-1999. Kalirajan and others (2001) explain that two important reasons for the slowdown are that there was no major breakthrough in developing new high-yielding varieties during the 1990s and there was a decline in the environmental quality of land which reduced the marginal productivity of the modern inputs. What could this mean in terms of the effectiveness of the policies of reduced protection to industry, a market determined exchange rate and the opening of the agricultural sector to foreign trade? First, although the reduction to protection of industry is substantial, there is reason to believe that the reduction was not necessarily sufficient to benefit the agricultural sector whose tariffs were also drastically reduced. Hence, the expected shift in resources to agriculture did not occur.

Second, is the apparent ineffectiveness of the market determined exchange rate in boosting exports. This is however not surprising as the exchange rate may not be a key factor determining agricultural export demand for India. In general, unlike manufacturing industries, agriculture did not benefit much from these two policies because the share of imported inputs in the value of agricultural production is small. It is likely that a change in the mindset and attitude of farmers has yet to take place and there are delays or hesitation in embracing India's openness.

Third, in opening up the agricultural sector to foreign trade, India has taken major steps towards trade liberalization since 1991, partly on its own initiative and partly from its commitments to WTO. Kalirajan and others (2001) provide a detailed review of these reform procedures. But why have the benefits from trade liberalization been slow to come?

One reason is that prospects for growth in agricultural exports depend partly on domestic policies and partly on the removal of protectionist policies pursued by developed countries such as Japan and members of the European Union (EU). An OECD report (1998) estimated that the producer equivalent subsidy in the OECD countries increased by US$ 9.3 billion from 1988 to 1993 and this subsidy as a percentage of the value of production in 1997 was 9 per cent in Australia, 20 per cent in Canada, 47 per cent in EU and 70 per cent in Japan.

These protectionist practices do not seem likely to come to an early end. An UNCTAD report (1999) noted that 29 member countries of the OECD spent an average of US$ 350 billion a year in agricultural support between 1996-98. Schumacher (2000) further reports that the EU provides product-specific trade distorting domestic support to at least 50 different agricultural products. The implication of these reports is that food exports from India may not show a large increase given the international environment and the still-existing restrictions on exports in the major importing markets based on the self-sufficiency argument and food security. Other macroeconomic factors, such as the recession in developed countries in 1996-98 as well as the 1997 South-East Asian financial crisis, have clouded the possibilities of increasing Indian exports. Another problem faced by Indian agricultural exporters is the protectionist measures in the form of non-trade barriers that developed countries use to restrict market access.

This is by tightening requirements of quality, testing and labeling, and anti-dumping and countervailing measures. For example, in May 1997, the EU banned marine products from India citing unhygienic processing conditions. The extra costs of meeting the standards required in export markets as well as costs associated with changes in the production mix and transactions associated with exports may well be discouraging Indian exporters. One existing problem of India's agricultural protection is the use of input subsidies. The general argument favouring this has been that it is necessary to encourage the use of particular inputs for production for various benefits. For India, Gulati and Sharma (1995) show that the input subsidy in per cent of GDP increased from 2.13 in the triennium ending 1982-1983 to 2.73 in the triennium ending 1992-1993. But the benefits of these subsidies have accrued to only certain classes of farmers in some regions cultivating irrigated crops.

Furthermore, highly subsidized prices of inputs such as irrigation water and electricity for pump sets have encouraged cultivation of water-intensive crops, over-use of water, ground water depletion/salinity and water logging in many areas. Subsidy for nitrogen fertilizer on the other hand has resulted in nitrogen phosphorous potassium imbalance and acted as a disincentive for use of the environmentally friendly organic manure. As a result, the linkage between food crops and non-food crops, which include fodder, has been reduced. These adverse consequences are a drain on the fiscal burden of central and state Governments.

Thus, if not properly monitored, input subsidies can be counterproductive and, in this context, protection to lower costs of production should be done selectively in the course of liberalization. In fact, Agenda 21 of the United Nations Conference on Environment and Development in 1992 stressed that there is a need for integration of environmental considerations in the pricing of natural and other resources in such a way that prices reflect social costs. Such a pricing policy will not only lead to a more efficient use of scarce resources but also result in subsidy reductions and improvements in environmental quality.

The money saved from the reduction of subsidies can be spent in the development of rural infrastructures, agricultural research, farmers' education and other forms of support for agriculture.

GROWTH AND PERFORMANCE OF AGRICULTURE

While the above analysis has provided a general view of the impact of economic reforms, this chaper examines agricultural growth and performance in the states of Bihar, Karnataka, Tamil Nadu and Punjab with their attendant policy implications. The yields for various crops in these states differ greatly. While Tamil Nadu had the highest yield in rice, oil seeds and sugarcane, Punjab enjoyed the highest yields in wheat, coarse cereals, pulses and food grains. Karnataka on the other hand is seen to do well in cotton and Bihar performed quite well in pulses and coarse cereals.

Further analysis and findings by Kalirajan and others (2001) show that Punjab had made remarkable achievements on the agricultural front while Bihar had remained stagnant in the last two decades, with Karnataka and Tamil Nadu showing moderate achievement. Clearly, differences in physical endowments, climatic conditions and institutional characteristics are some of the reasons for the varying productivity performance. Thus, having across the board economic reforms is likely to work less effectively than state-specific policy measures that enable each state's agricultural yields to reach their full potential. The comparative advantage of each state's agricultural production should be determined and with inter-state restrictions removed, total agricultural output would see a very significant increase. For example, Karnataka with less favourable soil and water resources should be given incentives to concentrate on agro-processed products and corporate agriculture in horticulture, floriculture and animal husbandry, or to undertake watershed development to help with dry land agriculture.

Many studies have indicated that with watershed areas, productivity growth has been mainly due to seed and fertilizer use. Thus, this state has to be given input subsidies for high yielding seed varieties but at the same time, the farmers need to be educated on the over use of chemical fertilizers. With Bihar, agricultural performance is problematic on many fronts.

First, although demographic pressure has increased and agricultural technology has improved, most of the uncultivated land is concentrated in southern Bihar, where irrigation facilities have not kept pace and the soil is of poor quality. Given the physiography of southern Bihar, wells are also unsuitable and thus the dominant mode of irrigation has been through tanks whose expansion and maintenance has been neglected.

Second, the infrastructural facilities of Bihar have been lagging as seen by the infrastructure development. Due to infrastructural bottlenecks, availability of modern goods and services has not increased or their supply remains costly or unreliable.

Third, agriculture in Bihar is dominated by small and marginal farmers and the prevalence of mass poverty is largely related to the backwardness of agriculture. Fourth and importantly, the state agricultural policies in Bihar are in dire need of review. The semi-feudal production condition still exists in rural areas and the ineffective protection of tenancy rights has hindered agricultural growth. The slow pace of land consolidation reflects inadequate financial outlays and a shortage of manpower. Kalirajan and others (2001) note that marketing and extension services in Bihar are also rather weak compared to the other states. Punjab on the other hand, was one of the few states which enjoyed the success of land reforms and the high priority of investment in rural infrastructure. Also, the irrigation base of the small and medium sized farms was comparable to that of large farms. In addition, the Punjab Agricultural University at Ludhiana contributed to the development of new seed varieties.

However, there are clear signs of a decline in crop yields since the 1990s and this has been associated with the increasing use of fertilizers and excessive water use which have increased the unit cost of production as a result of declining soil quality. Hence, care is needed when providing further input subsidies in fertilizer and water use. Another related fact is the steep increase in wages in Punjab and in the absence of productivity increases, the cost increase has affected the profitability of farmers. With Tamil Nadu, the main crop has been rice as this state is blessed with two monsoons. But from 1992-1997, there has been a steady decline in the areas irrigated by canals and an increase in well-irrigated areas while the use of tanks remains an unreliable source of irrigation. However, major improvements in about 10 rice varieties released in the early 1990s can be expected to improve productivity growth in rice production although pests and diseases as well as imbalance in the use of fertilizers are major constraints. Thus Tamil Nadu could do with subsidies of pesticides and farmers should be educated on the more effective use of fertilizers to obtain high yields. Interestingly, the cropping pattern of late has shown increasing substitution of food crops by commercial crops but there is concern that the benefits will reach farmers only with the development of adequate infrastructure such as roads and markets. However, that Tamil Nadu has a higher index than the all India average of infrastructure.

Challenges of Globalization

It is important to realize that globalization poses many challenges to a developing country like India, which had relied on a state directed and regulated policy regime for more than four decades. In moving to a more open, market-based economy there are many transitional problems that the country has to manage. The Government must play a pro-active role in facilitating the globalization process so that the opportunity sets for the economic agents are widened and the adverse effects of globalization are minimized. The Indian Government must also prepare the necessary information base and develop its capacity to articulate India's concerns and policy trade-offs in the international forums for multilateral trade and environmental negotiations.

In addition, the Government should embark on an extensive programme to educate farmers on the need to meet the standards required in the export markets. In fact, India needs to seek technical assistance in creating the capacity for meeting such standards and to consider watershed developments for environmental considerations. Equally important is the need to disseminate information about possible export markets to farmers, so that market access is achieved at minimum cost. Given the requisite information about markets and profitability, the likelihood of farmers investing in post-harvest and processing technologies and storage and efficient transportation arrangements as well as developing supporting infrastructure is very high.

Although the brave and bold move by India to reduce the tariff rate for agricultural products from 113 per cent in 1990-1991 to 26 per cent in 1997-1998 deserves to be applauded, the question of whether India is ready to compete in world markets remains to be seen. The infant industry argument may still hold for India to shield itself from external competition but one can easily question the length of time that is required to that end. Also, a delay in opening up to foreign trade has the danger that local producers may become too complacent and never be ready for competition.

As India opens up externally, it is also expected to face vulnerability in the wider international price fluctuations and thus Acharya (1998) claims that a minimum price support scheme is important. These prices can also act as a signal to adopt modern inputs and invest in yield-raising infrastructure for increasing production. For instance, keeping basic staple food grains at reasonable prices would induce farmers to switch over to high value crops. However, during the 1990s, Kalirajan and others (2001) shows that procurement prices especially for rice and wheat have been increasing faster than the general price level. Such high prices along with guaranteed purchases by the Food Corporation of India have pushed up market prices. These higher prices are partly responsible for the large buffer stocks with the Food Corporation. If this trend continues, India's comparative advantage will be eroded.

With openness and high price instability, unstable export revenue can also be expected. One way of reducing such risk is for India to diversify her agricultural exports. For example, since 1990, even in commodities such as tea, coffee, cocoa and spices, where India is supposed to have a comparative advantage international prices have been unstable. Besides increasing the type of exports to obtain more export revenue, India should also seriously consider exporting more value added agricultural products through agro-processing such as processed vegetables, fruits, fish and meat products given that export or even local demand for basic agricultural products would decline as incomes rise. The move to higher value added activities within the agricultural sector also spells greater opportunities for industrialization and vice versa as borne by Kalirajan and Shand's (1997) findings of a bi-directional relationship between agriculture and industry for most Indian states. On the other hand, Sivakumar and others (1999) establish empirical evidence of high forward linkages of agriculture due to the presence of agro-industries while Satyasai and Viswanathan (1999) show the significance of the spillover effects to the industrial sector via the intensive use of purchased inputs in the agricultural sector.

The lack or slow pace of internal or domestic liberalization is also seen to hinder the possible gains from external or trade liberalization. For example, although central zoning restrictions have been abolished, state government restrictions on inter-state and even inter-district restrictions on marketing and movement of goods still exist in many cases. This interferes with the benefits from crop specialization and economies of scale arising from comparative advantage. The land market is another example of distortion whereby land ceilings exist preventing the operation of large-sized farms. This has led to the emergence of a large number of small economically unviable land holdings. The easy leasing of land should be permitted with assurance of resumption. Yet another problem lies with the insufficiency of credit to agriculture. From 1995-1996, the Rural Infrastructure Development Fund was set up to allocate funds for the completion of projects and the government has committed itself to strengthening the cooperative credit structure through substantial refinancing and restructuring of the Regional Rural Banks. However, as mentioned earlier, due to varying institutional factors in the Indian states, these domestic reforms can be expected to yield quite different results.

Although India missed the opportunity to open up two decades ago, its attempts to do so now must be regarded as better late than never. Others such as Desai (1999) observe that, "the logic of the global economy as well as India's interests dictate that India become proactive in its liberalization policies. India must liberalize not because it has no choice but because it is the best choice".

His lament that India has adopted a 'victim mentality' when it really needs to adopt a 'winner mentality' has become less of a concern as over time, India has shown commitment to stay on the bandwagon of globalization. Having realized that globalization is a necessary but not a sufficient condition for high growth production, India has undertaken economic reforms, both internal and external. However, it must be ensured that these reforms are synchronized so that the pace of both reforms is set right in order to work hand in hand to promote agricultural productivity growth.

Thus, training the farmers and educating them appropriately to change their mindset and reorienting them to take up new activities or adopt foreign technology is of utmost importance. In this context, it is necessary to involve non-governmental organizations in training and mobilizing the rural poor to face the challenge of liberalization.

Also, with domestic economic reforms, more care needs to be exercised to draw up state-specific liberalization measures to maximize their benefits. Lastly, in the implementation of these reforms for successful globalization, one crucial element, not entirely within control is the need for good governance and stability in the political and economic environment. Political leaders who are the ultimate decision makers in these matters need to examine their own role dispassionately. It is quite apparent that at this relatively early stage, there is little observable evidence of gains to India's agricultural performance after opening up. However, there could easily be benefits that have not yet surfaced, or are yet to be identified and perhaps too difficult or intangible to measure.

Whatever the case, it is highly likely that it is too soon to assess the full impact of globalization and economic reforms. Furthermore, the process of liberalization has been gradual and remains incomplete. For example, the complete removal of quantitative restrictions after March 2001 will have provided an opportunity for Indian farmers to tap world markets and, if they are successful, results should start to become evident soon. Export promotion via the development of export and trading houses as well as effective liberalizing export promotion zone schemes for agriculture are fairly recent measures and only time will tell as to how effective these measures are. Other possibilities such as agro-industry parks for promoting exports are also in the pipeline.

PRODUCTIVITY GAINS FROM GLOBALIZATION AND ECONOMIC REFORMS

In the wake of India's efforts towards globalization and economic reforms, the expected benefits of total factor productivity (TFP) growth can be represented using the production frontier. The production frontier traces out the maximum output obtainable from the use of inputs.

Opportunities from globalization and economic reforms can lead to:

- shift from A to B due to technical efficiency
- shift from B to C on existing frontier due to input growth
- upward shift from C to D due to technological progress

Which constitute various sources of TFP growth, can be linked with trade gains. The movement from A to B led by technical efficiency allows increases in output when inputs and technology are used to their fullest potential to obtain the greatest yield.

Given that India has been involved in agricultural production for so long, there would be learning-by-doing gains that can help boost production given the expected increase in demand as India opens up. The increased production would enable a better utilization of inputs, especially that of advanced capital technology. The reduction in the tariff rate for agricultural products from 113 per cent in 1990-1991 to 26 per cent in 1997-1998 is also expected to motivate local producers into rethinking their production techniques and efficiently utilizing the inputs and technology to keep costs of production down in order to remain competitive. The optimum or efficient use of land and water resources would then allow agriculture to respond to the demand for other products such as horticulture and livestock which is expected to increase following a rising trend in the per capita incomes of both rural and urban groups.

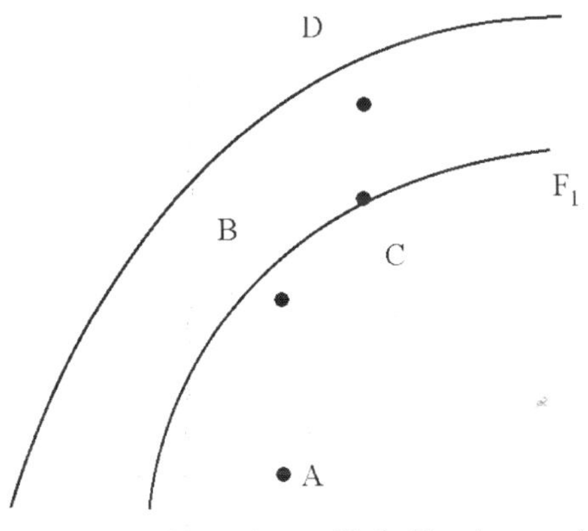

Fig. Total Factor Productivity Gains from Globalization and Economic Reforms

The move from an overvalued exchange rate to that of a market determined rate would also make agricultural exports cheaper and hence boost exports. The new trading opportunities would necessitate an increased use in the quantity of inputs to boost output and this allows for the movement from B to C along the existing production possibility frontier. Increased exports would bring about economies of scale and as Verdoon's law states, output growth would lead to productivity growth.

The scale of output under increased exports would justify the huge fixed costs underlying technologically advanced equipment and hence increase incentives to adopt high quality inputs. The use of such inputs would result in technological progress and this is represented by the shift from C to D. The reduction in tariff rates in industry from 1990-1991 to 1997-1998 range from 153 per cent to 25 per cent for consumer goods, 77 per cent to 18 per cent for intermediate goods and 97 per cent to 24 per cent for capital goods.

This means that farmers now have relatively cheaper access to imported new technology and better capital equipment as well as the option of adopting better farming techniques and this should lead to technological progress. In particular, the development of agro-processing as an instrument for agricultural and rural modernization will bring benefits, given its capital-intensive and technology-intensive nature.

Lower duty rates on plastics and metals also lower costs of packaging. These forms of cost efficiency should allow competitive pricing of products. In addition, external competition can be expected to motivate local producers into the production of improved quality intermediate inputs for agriculture.

The importance of technology in agricultural development was first demonstrated in the 1970s with impressive growth in yields following the introduction of new wheat and rice varieties. But this technology was limited to areas of assured irrigation as the new seeds also required heavy inputs of fertilizers and pesticides for optimal results. However, the potential for further extending this technology is not yet exhausted as there is scope for expanding irrigation further and improving the quality of irrigation in many areas. For further technological progress, genetic engineering and the biotechnology revolution provides a prospect of developing new varieties that can flourish with less dependence on water and chemical inputs. Such reduced dependence upon chemical fertilizers and pesticides is also desirable because of environmental considerations, which are an increasing concern.

It must however be acknowledged that the link between trade liberalization and productivity growth is two-way as they both feed on each other. The productivity gains can be obtained from openness but to benefit from openness via increased demand for exports, agricultural products need to be priced competitively. In other words, productivity growth is necessary to lower the costs of production.

INDIA'S ECONOMY REFORM IN AGRICULTURAL SECTOR

In the early 1950s, half of India's GDP came from the agricultural sector. By 1995, that contribution was halved again to about 25 per cent. As would be expected of virtually all countries in the process of development, India's agricultural sector's share has declined consistently over time as seen in the table below.

Table : Share of agricultural output in India's GDP

Year	1950/51	1965	1976	1985	1991	1999
Percentage share	52.2	43.6	37.4	32.8	28.3	24.4

In the last five decades, the Government's objectives in agricultural policy and the instruments used to realize the objectives have changed from time to time, depending on both internal and external factors. Agricultural policies at the sectoral level can be further divided into supply side and demand side policies. The former include those relating to land reform and land use, development and diffusion of new technologies, public investment in irrigation and rural infrastructure and agricultural price supports. The demand side policies on the other hand, include state interventions in agricultural markets as well as operation of public distribution systems. Such policies also have macro effects in terms of their impact on government budgets. Macro level policies include policies to strengthen agricultural and non-agricultural sector linkages and industrial policies that affect input supplies to agriculture and the supply of agricultural materials.

During the pre-green revolution period, from independence to 1964-1965, the agricultural sector grew at annual average of 2.7 per cent. This period saw a major policy thrust towards land reform and the development of irrigation. With the green revolution period from the mid-1960s to 1991, the agricultural sector grew at 3.2 per cent during 1965-1966 to 1975-1976, and at 3.1 per cent during 1976-1977 to 1991-1992. Acharya (1998) explains that the policy package for this period was substantial and consisted of:

(a) introduction of high-yielding varieties of wheat and rice by strengthening agricultural research and extension services,
(b) measures to increase the supply of agricultural inputs such as chemical fertilizers and pesticides,
(c) expansion of major and minor irrigation facilities,
(d) announcement of minimum support prices for major crops, government procurement of cereals for building buffer stocks and to meet public distribution needs, and
(e) the provision of agricultural credit on a priority basis. This period also witnessed a number of market intervention measures by the central and state Governments.

The promotional measures relate to the development and regulation of primary markets in the nature of physical and institutional infrastructure at the first contact point for farmers to sell their surplus products.

Acharya (1998) also notes that the rate of growth of productivity per hectare of all crops taken together increased from 2.07 per cent in the decade ending 1985-1986 to 2.51 per cent per annum during the decade ending 1994-1995. Similar evidence of an increase in yields, a partial measure of productivity gains given by output per unit of land area is seen below for various crops.

Although productivity gains were sustained in the 1990s after the liberalization process began, the yield rates for most of the agricultural products in India are far below comparable rates in a number of other countries. Except for sugarcane, tea, coffee and jute, India's yields are lower than the world average. It should be noted that India is ranked second both in area and output for sugarcane production and is the largest producer of tea and jute in the world. Although India is doing quite well in wheat production, the average yields in the Netherlands and Ireland are more than three times India's yield rates. In all other major crops, India's productivity performance seems to lag behind others.

Table. Yield for various crops (kg/ha)

	1950/51	1960/61	1970/71	1980/81	1990/91	1995/96	1998/99
Rice	668	1 013	1 123	1 336	1 740	1 855	1 905
Wheat	663	851	1 307	1 630	2 281	2 483	2 596
Coarse cereals	408	528	665	695	900	941	1 035
Pulses	441	539	524	473	578	552	661
Food grains	522	710	872	1 023	1 380	1 499	1 611
Oil seeds	481	507	579	532	771	851	948
Cotton	88	125	106	152	225	246	240
Sugarcane	33 422	45 549	48 322	57 844	65 395	68 369	69 288

Why Globalize?

Globalization in the context of agriculture can be best discussed in the context of three components – improvement of productive efficiency by ensuring the convergence of potential and realized output, increase in agricultural exports and value added activities using agricultural produce, and finally, improved access to domestic and international markets that are either tightly regulated or are overly protected.

These components are linked in various ways. For example, productive efficiency would enhance value added activities in agriculture through agro-processing and exports of agricultural and agro-based products. These activities in turn would increase income and employment in the industrial processing sector. Thus globalizing agriculture has the potential to transform subsistence agriculture to commercialized agriculture and to improve the living conditions of the rural community.

Ahluwalia (1996) explains that this indirectly requires an improvement in agricultural growth from between 2 and 3 per cent in the past to about 4 per cent per year. Although initially, with respect to agriculture, there was no major policy reform package in the 1990s, it was however anticipated that the opening up of the agricultural sector to foreign trade, the move to a market determined exchange rate and reduction of protection for industry would, over time, benefit the agricultural sector.

Manmohan Singh (1995), the then Finance Minister, in his inaugural address at the 54th Annual Conference of the Indian Society of Agricultural Economics, brought to notice that a policy of heavy protection of the industrial sector operated to the disadvantage of the agricultural sector when industrial prices were raised relative to world prices and thus the profitability of investing in industry was raised relative to agriculture. This would lead to a shift of resources from agriculture to industry. A policy of heavy industrial protection also led to an appreciation of the exchange rate. Ahluwalia (1996) noted that over-valuation of the exchange rate (before the Indian rupee was devalued by 18 per cent in two phases starting in July 1991) discouraged agricultural exports more than industrial exports because Indian industrial policy had sought to offset the constraints faced by industries via a system of export incentives for market support. Agricultural exports on the other hand were denied any such incentives as they did not use imported inputs.

Ahluwalia (1996) argued that in the past, the agricultural sector was negatively protected because of the above two reasons and the fact that farmers were denied access to the world markets due to trade barriers. Exports of plantation crops and a few commercial crops were free from export restriction but exports of essential commodities, particularly food products, were subject to bans, quotas and other restrictions. Interestingly, Kruger and others (1991) showed that while many developed countries continue to protect agriculture, developing countries do not do so.

However, no formal attempt or theoretical framework has yet been used to assess the extent of negative protection in Indian agriculture. The implementation of economic reform in the Indian agricultural sector has been a gradual process. These include an 87 per cent cut in tariff on agricultural products, sustenance of high-yield crop varieties, removal of minimum export price on selected agricultural products, a lift on quantity restrictions on the export of some crops and various land reforms related to tenancy rights and land ceilings.

THE SHORT TERM: ECONOMIES OF CAPACITY UTILIZATION

Efficiency in the appropriation, storage, distribution and use of water for agriculture. In all cases but one these are *output efficiency* measures, having a crop output or crop sales measure as the numerator and a measure of irrigation water supply as the denominator.

Of the first seven efficiency measures, only one is a supply-side ratio. This is *E*2 -net irrigation requirement divided by the base supply. *Supply-side efficiency* in terms of the supply cost per cubic metre of water delivered to the farmer's field. It is also important to remember that irrigation supply in cubic metres can be measured at many points. An irrigation engineer writes:

In big systems there are very substantial losses before the water gets to the irrigation district boundary; there are more between that and the farm boundary and, where farms are big and open channel distribution systems in use, more losses before the water gets to the field. Finally, for the purpose of comparing alternatives for on-farm delivery, one should measure the supply both at the field boundary and at the plant.

The two measures most often used in this text are *S*b, base supply at the start of the irrigation cycle and *S*f, the volume delivered to the farmer's fields. The most important concepts in the irrigation economist's tool-kit to explore supply-side efficiency are *cost functions*.

These quantitative relationships describe the cost of supplying output in any time-period at each scale of output from zero units up to the system's theoretical capacity. The ability to shift from one delivery level of irrigation water up to a higher level clearly depends on the time available to make the change. It was Alfred Marshall, one of the greatest British economists, who introduced periodization into the analysis of supply and demand. He distinguished between the short term and the long term. In its application to the irrigation cycle, this simple, twofold distinction can be represented in the following way: in the short term, increased daily output is possible through operational changes or by organizational innovations demanding new procedures, both of which are relatively straightforward in their introduction; in the long term, additional infrastructure is required either in new projects or for the expansion of existing capacity, with associated financial, resource and organizational planning.

In conventional analysis it is possible, for either the short or long term, to calculate the *total cost* of supply at each level of output per day. *Average cost* at each level of output can be derived from this by dividing the total cost by the number of units produced.

The *marginal cost* of the *n*th unit of output is defined as the difference in total cost between producing *n* and *(n-1)* units. A related but alternative exposition is set out below. It has the advantage that the terms refer to engineering categories, making it easier for non-economists to grasp quickly. It is also more appropriate than the conventional analysis, since it is a better representation of the way irrigation institutions actually think about their costs. Expenditures are defined using three categories-the parallel with farming family and agribusiness costs:

- Prime costs are defined as those used up in the daily supply of irrigation services and consist of the salaries and wages of the workforce, the costs of power, materials, spare parts and other consumables such as bought-in specialist inputs.
- Overhead costs of production are non-prime costs and are often assumed to be invariant with the level of supply. In the simplest case, where capital expenditure on land, infrastructure and plant has been funded 100 per cent by loans, this element of annual overhead costs would be set equal to the annual interest payable and the principal repayable on the debt incurred. Where capital investment is funded, at least in part, from an irrigation authority's surplus of water charges over prime costs, overhead costs include the amortization charges necessary in the long run to contribute to the rehabilitation and replacement of elements of the system infrastructure. The rent of land and buildings and the leasing costs of capital equipment such as vehicles, are also included in overheads.
- Total costs per year are equal to prime plus overhead costs.

For each level of output, one can calculate overhead cost per unit of output and this function, called average overhead cost, is represented for a hypothetical case, *illustrating the short-term situation.* Average overhead costs fall (at a decelerating rate) as the unchanged level of cost is divided by evergreater levels of production. So in the short term, supply-side efficiency in terms of the overhead costs of irrigation water supply, *i.e.* the supply costs of the headworks and networks infrastructure per unit of water delivered, is achieved by operating the irrigation system at its full capacity. In other words, at scales of delivery less than full capacity, the average cost of the appropriation, storage and distribution infrastructure necessary to supply irrigation water rises exponentially.

Where an irrigation system is built to meet the peak period of demand-which may be only 10 days long-the system is operating at below capacity for most of the irrigation season. Alternatives are to design a system that does *not* meet peak demand or to use volumetric pricing to suppress the peak demand for water. In this respect Carruthers and Clark showed an interesting link between surface-water and groundwater abstraction;

Tubewells may be used where surface canal water is at present supplied to augment basic surface supplies and increase cropping intensity. Augmentation to meet relatively short periods of peak demand may be particularly valuable. The most profitable cropping pattern is unlikely to have an even profile of water demand over the year and a canal system is limited in its capacity to cope with peaks.

THE LONG TERM: ECONOMIES OF SCALE

We now switch to the long term. Again our interest is in how average overhead costs vary with output scale. In this case output is increased not by raising the volume of irrigation water supplied towards 100 per cent utilization of *a fixed* infrastructural capacity, but by *adding* to the infrastructural capacity through specified engineering works. We can say economies of scale exist where higher levels of output are associated with lower overhead cost per unit of irrigation water supplied.

For the long term, where output rises because of capacity increases resulting from the investment process, the best approach in understanding scale economies is through *ex ante* calculations, *i.e.* a view of future possibilities not yet realized-*a planning approach.* The starting point is a real catchment in a specific year, with all the infrastructural investment in irrigation services inherited from the past in place. The interest here is in average overhead cost at each alternative value of a set of supply volumes derived from additions to capacity, from a minimal increment through to any upper limit that one believes to be appropriate. Note that the curve does not consist of a set of points representing how average overhead cost changes as the scale of production expands (or contracts) over time. It is not an evolutionary function.

What it represents, for each output level, is the average overhead cost if, in a single leap, capacity were expanded up to that level. It is referred to as a long-term supply function not because the figure represents economies of scale with the passage of time but because the long term is the conventional time-scale over which additions to capacity can take place, in contrast to the operational changes of the short term. This is worth restating. Average overhead cost variations with output in the short run represent the cost-output relationship as management in real-time shifts output upwards or downwards from one level to another.

Average overhead cost variations with output in the long run represent the cost-output relationship for each of a series of separate, mutually exclusive choices. Here it is logically impossible for there to be real-time shifts of output upwards or downward from one level to another. So, in the short term, shifts over time in output levels are represented; but in the long term, mutually exclusive choices between future additions to capacity and to output are represented. The hypothetical function clear economies of scale up to 140 units of output per day. However, here, diseconomies of scale appear above 150 units per day. This cost function is quadratic in nature, *i.e.* well represented by an equation of the form:

$$\text{AOC} = L^2 + bL + c$$

where AOC is average overhead cost and L is the output quantity the equation is:

$$\text{AOC} = 0.004L^2 - 0.998L + 93.5$$

How can one explain long-term economies of scale in irrigation service provision? It is the measure of what economists widely refer to as the *indivisibility* of infrastructural provision. This neatly expresses the idea that, to secure even modest levels of output, major works are necessary-a canal is a good example-and these are negligibly less costly than works that secure more substantial output levels. As a result, AOC falls markedly. The existence of significant scale economies at low- to medium-output levels has a most important social effect. It implies that irrigation service supply constitutes what is widely called a natural monopoly. The same can hold true in infrastructural provision for fresh and wastewater services in general, as well as for transport, energy supply and telecommunications. The proposition that long-term economies of scale exist in the supply of irrigation services should be used more as a working hypothesis than as an accepted fact. Moreover, one can expect variation in the existence of such economies between the infrastructural categories of appropriation, storage and distribution.

With respect to appropriation headworks, the existence or not of economies surely varies between the five base flows of the irrigation cycle. *Rainfall collection* using a harvesting surface and a cistern for storage can work well at the small scale and is used in many parts of the world by farming families for small plot cultivation, particularly for their subsistence needs. We saw such an example in the case study of the small agribusiness in Cornwall, where rainfall collection contributed one-sixth of the farm's irrigation requirements.

Irrigation water supplied by *the reuse of domestic and industrial wastewater* certainly can be expected to exhibit long-term scale economies in respect of any wastewater treatment prior to reuse. However, if the wastewater treatment plant's (WWTP) cost is accounted for as part of the domestic and industrial supply system, the average overhead cost of reuse water appropriation is zero.

The reuse in farming of *agricultural drainage water* by means of pumps set in drainage channels seems to operate very effectively at small scales, so no long-term scale economies can be expected here.

Surface-water and groundwater abstraction takes place on a wide variety of scales as we have already seen in the Tamil Nadu, Curu Valley and Cornwall case studies. I do not believe that one can expect to find a consistent pattern of long-term scale economies here. The variation in technology is phenomenal, from the treadle pump up to major infrastructures for river diversion. An extensive review of the use of the treadle pump in Africa shows how much more effective it is than traditional rope-and-bucket irrigation systems. In Bangladesh more than 500,000 of these low-cost devices are in daily use. Even where scale economies exist for one component of the irrigation cycle, such as a river barrage like that on the Bhavani River, distribution costs may be greater across the large area in command.

With the exception of springs and of surface-water diversion through civil engineering structures large or small, virtually all groundwater and surface-water abstraction is carried out by pumping. So the possible existence of long-term scale economies in pumps and their power units needs to be reviewed. Positive displacement pumps such as the piston pump, the rotary pump, the air-lift pump and the Archimedean screw are convenient for small discharges. Of the roto-dynamic pumps (centrifugal, axial and mixed flow) centrifugal pumps are the most widely used. They are ideally suited to small discharges at high pressures but in fact are very versatile. Axial pumps are very efficient at lifting large volumes of water at low pressure; they tend to be used for large pumping works because of the expense of their drive shaft and bearings. For some middle-range head and discharge requirements, mixed flow pumps (combining the centrifugal and axial flow principles) perform best. Centrifugal pumps can also be operated in series when extra head is required.

Pumps and their power units convert fuel energy such as diesel oil into useful water energy. One of the interests of hydraulics is how well the power from the engine or motor is converted into useful water power in the pump. This is given as a measure known as power efficiency. In the absence of friction, efficiency would be 100 per cent but there are always friction losses in all the components of both the pump and its engine. Economies of scale *do* exist here. As Melvyn Kay writes (1998:230): 'Smaller pumps are usually less efficient than larger ones because there is more friction to overcome relative to their size.' Force-mode in contrast with suction-mode pump technologies are more expensive but give access to groundwater at greater depth. Here, in particular, the larger investment leads purchasers to give greater consideration to pump and engine performance, reliability, repair facilities and spare parts availability.

With respect to power units, electric motors are very efficient in energy use and can be employed to drive all sizes and types of pump. Where access to electricity is not easy or is unreliable, petrol or diesel engines are used. Diesel engines are heavier and more expensive to purchase but they are more robust than petrol engines, have a much longer working life and are cheaper to run.

Turning now to storage head works, strictly in terms of their engineering economics, dams certainly offer economies of scale. For example, large dams in terms of storage capacity on balance have a greater depth of water and therefore lose less of their water to evaporation. Again, as Melvyn Kay points out in his excellent introduction to hydraulics, the thickness and therefore the cost of a dam wall's construction is determined by the head of water it contains, not by the volume of water it stores. This is known as the dam paradox. In general, the ratio of the number of cubic metres of water stored to each cubic metre of wall increases with dam size. The empirical relation between dam size and capital cost for each thousand cubic metres stored strongly confirms the existence of scale economies.

The costs of storage are also strongly affected by site conditions: in broad valleys with poor foundation conditions, costs will be high. An ideal site is a narrow gorge broadening to a large valley up-stream with good, solid rock conditions and no risk of earthquake activity. In the case of storage, as with other infrastructural works, the best sites tend to be developed first.

Max Neutze stresses that unambiguous scale economies exist for the *networks* of land drainage, flood control and fresh and wastewater services, including irrigation service supply. The cost of pipes increases roughly in proportion to their length and diameter, but their capacity increases in proportion to their diameter to the power of about 2.6. This is because the cross-sectional area increases as the square of the diameter and because friction between the water and the pipe decreases with size. Indivisibility, Neutze argues, is a characteristic of networks. On the subject of networks, an advantage of boreholes over surface-water surfaces is that they can be sited adjacent to the areas to be irrigated, thus reducing distribution overheads. Stressed that the long-term cost function deals with additions to capacity 'in a single leap', that evolutionary development is not in question here. However, when engineers are preparing plans many years ahead, it may be necessary to plot an evolutionary path. In the case of a 10-year water strategy, for example, it may make sense to consider two rounds of investment.

Here one is concerned not with a once-for-all heave to arrive at a given output, as with the long-term supply function, but with a first investment commitment followed by a second round of construction, for example, as an agricultural area's demand expands over time. Average overhead cost for the first of the two investment rounds in our example. But the AOC of the second round is not so defined, since the long-term cost function will have shifted as a result of the prior introduction of the first-round infrastructures.

Here is an example. Suppose that the stock of water in a reservoir for an irrigation district when it is fully developed will need to be 100 units. The single-round solution might simply be to build a 100-unit facility. A two-round solution might be to construct a 75-unit reservoir initially and then to add 25 units more in a second stage. The single-throw approach may have the advantage of providing a more cost-effective solution when the district is fully in production but at the cost of underutilized capacity during the first phase of the district's development.

But note, too, that the second-round addition of 25 units might be achieved either by a free-standing 25-unit reservoir or by deepening the 75-unit facility. With a double-throw, the first investment round changes the options available when the second round arrives. So, we can define the *indivisibility* of water infrastructures as what makes a single 20-m dam cheaper than two dams each of 10 m-a cross-section comparison. The *lumpiness* of water infrastructures is what makes a 10-m dam built now impose a higher construction cost for an additional 10-m capacity a year later-a time-series comparison. Lumpiness particularly characterizes the irrigation networks of pipes and canals. Lumpiness occurs because land prices will have been pushed up by first-round infrastructural construction, because pipe and canal installation is more expensive when it necessitates digging up an existing distributional infrastructure and because second-round investment is more disruptive than that on virgin sites.

If the range of second-round options are part of a long-term infrastructural plan in designing the first-round facility, the 25-unit expansion is likely to be achieved more cost-effectively than without such planning. This is the basis of Neutze's argument that the durability, specialization, immobility, indivisibility and lumpiness of infrastructural provision provide a strong economic argument for land-use planning and the demand management of growth. Note that groundwater abstraction is far less lumpy than surface-water abstraction. Tubewell construction can be phased in with demand, unlike canal systems and dams. The attractions of short-term economies of capacity utilization contrasted with the potential advantages of long-term planning and the link between prime and overhead costs, are illustrated by Kay:

Pipe sizes are often selected using a discharge based on present water demands and little thought is given to how this might change in the future. Also there is always a temptation to select small pipes to satisfy current demand simply because they are less expensive than larger ones. These two factors can lead to trouble in the future. If demand increases and higher discharges are required from the same pipe, the energy losses will rise sharply and so a lot more energy is needed to run the system.

As an example, a 200 mm diameter pumped pipeline, 500 m long supplies a (discharge) of 50 litre/s. Several years later the demand doubles to 100 litres/s. This increases the velocity in the pipe from 1.67 to 3.34 m/s (*i.e.* it doubles) and the energy loss rises from 5 m to 20 m (*i.e.* a four-fold increase). This increase in head loss plus the extra flow means that eight times more energy is needed to operate the system and extra pumps may be needed to provide the extra power required. A little extra thought at the planning stage and a little more investment at the beginning could save a lot of extra pumping costs later.

Note that Kay does not envisage adding a second 200-mm pipeline. The correct choice trading off short-term economies of utilization against lumpiness and long-term scale economies needs project evaluation over the life of the scheme. Infrastructural efficiency is not restricted to questions of scale economies. The cost and performance of different conveyance systems in Bangladesh in 1998. Capital costs per metre vary enormously. At the two extremes, buried asbestos pipes cost twenty-three times as much as high-density polyethylene (HDPE) surface piping-the distributional choice at Keveral Farm. There is little or no correlation between capital cost and either maintenance cost or distributional efficiency.

But the capital and maintenance costs of conveyance systems between the point of supply appropriation and the irrigated fields are not the only determinants of farmers' choices. Other factors of great importance are the *availability* of a distribution technology at the time of infrastructural construction, the *credit* provided for alternative technologies, the *permeability* of local soils and the hydraulic *head* available for water distribution, van Bentum and Smout (1994), for example, point out that soils that are sandy or loamy favour buried pipe systems. Scarce supply combined with low hydraulic head favours lined channels. Compacted earth channels score best when soils are neither sandy nor loamy, where water is not scarce and where low head disfavours buried or partially buried pipe systems.

In hilly areas the cost of channels is increased because their alignment must follow the land's contours to create a gentle downward slope for the flow. A more direct but steeper route would bring higher flow velocity causing erosion and serious damage to the channels. Pipes can be used in any kind of terrain, can take a direct route and the resulting high water velocity brings no risk of soil erosion. Channels are particularly attractive options for conveying large quantities of water over relatively flat land such as in large irrigation systems on river floodplains.

IMPACT OF ECONOMIC REFORMS ON AGRICULTURE IN INDIA

The serious foreign-exchange crisis in 1990 led to a number of well-publicized economic reforms in the early 1990s dealing with trade, industrial licensing, and privatization. The reforms had an impact on the agricultural sector through the central government's effort to withdraw the fertilizer subsidy and place greater emphasis on agricultural exports. The cut in the fertilizer subsidy was a result of the government's commitment to reduce New Delhi's fiscal deficit by removing grants and subsidies from the budget. The government action led to a reduction in the use of chemical fèrtilizers and protests by farmers and opposition political parties. The government was forced to continue the subsidies but at a somewhat lower level.

New import and export policies aim at enhancing export capabilities of the agricultural sector by increasing productivity and promoting modernization and competitiveness.

The measures to facilitate export growth include allowing the import of capital for the agricultural sector, reducing the list of agricultural products that cannot be exported, and removing the minimum export price from a number of products. Agricultural exports increased by 30 percent in FY 1991 and 14 percent in FY 1992 in terms of rupee value, but declined by 8 percent from FY 1990 to FY 1992 in United States dollar terms because of the devaluation of the rupee in 1991. In the mid-1990s, it was expected that agriculture would continue to be the most important sector of the economy for the rest of the decade in terms of the proportion of GDP.

However, even when it is not the sector providing the largest share of GDP, the importance of agriculture is not likely to diminish because of its critical role in providing food, wage goods, employment, and raw materials to industries.

Despite their preoccupation with industrial development, India's planners and policy makers have had to acknowledge the critical role of agriculture in the early 1990s by changing basic policy. The gains in agricultural production should not lead to complacency, however. Continuing increases in productivity, developing allied activities in rural areas, and building infrastructure in rural areas are essential if India is to continue to be self-reliant in food and agricultural products and provide a modest surplus for exports.

AGRICULTURE AND RURAL DEVELOPMENTS

From a nation dependent on food imports to feed its population, India today is not only self—sufficient in grain production, but also has a substantial reserve.

The progress made by agriculture in the last four decades has been one of the biggest success stories of free India. Agriculture and allied activities constitute the single largest contributor to the Gross Domestic Product, almost 33 per cent of it. Agriculture is the means of livelihood of about two—thirds of the work force in the country.

This increase in agricultural production has been brought about by bringing additional area under cultivation, extension of irrigation facilities, the use of improved high yielding variety of seeds, better techniques evolved through agricultural research, water management, and plant protection through judicious use of fertilizers, pesticides and cropping practices.

Crops

The 1970s saw a multi-fold increase in wheat production that heralded the Green Revolution. In the next decade rice production rose significantly; in 1995-96, rice production was 79.6 million tons. Total grain production crossed 191 million tons in 1994-95, a big leap from 51 million tons in 1950-5 1. During the Seventh Plan, the average grain production was 155 million tons, 17 million tons more than the Sixth Plan average.

To carry improved technologies to farmers, a National Pulse Development Programme, covering 13 states, was launched in 1986. The Special Food Production Programme augmented efforts to boost pulse production further. In 1995-96, pulse production was 13.2 million tons. With some States offering more than the statutory minimum price, sugarcane production also received a boost, in 1995-96 a record 283.0 million tons was registered.

Irrigation

As efforts continued to increase the irrigation potential in the country, the last 40 years saw the gross irrigated area reach 8~ million hectares. Flood forecasting has become an important activity over the years. Over 500 hydrological stations collect and transmit data through 400 wireless stations for issuing forecasts for 157 sites. About 5000 forecasts are issued in a year with 94 per cent accuracy. The country also receives international support, with the World Bank as a primary source, for developing the water resources. International cooperation is also envisaged in setting up a National Centre for Information on Water and Power.

As there is a broad seismic belt in the country, particularly along the Himalayan, the Kutch region and ports of Maharashtra, a scheme is being evolved to collect all data on seismic activity at various dam sites.

Fertilizers

The fertilizer industry in India has grown tremendously in the last 30 years. The Government is keen to see that fertilizer reaches the farmers in the remote and hilly areas. It has been decided to decontrol the prices, distribution and movement of phosphatic and potassic fertilizers. Steps have been taken to ensure an increase in the supply of non-chemical fertilizers at reasonable prices. There are 53 fertilizer quality control laboratories in the country. Since bio—fertilisers are regarded as an effective, cheap and renewable supplement to chemical fertilizers, the Government is implementing a National Project on Development and Use of Bio-fertilisers. Under this scheme, one national and six regional centers for organizing training, demonstrating programmes and quality testing of bio-fertilisers has been taken up.

It was a challenging decision of the Government to take Bombay High gas through a 1,700-km pipeline to feed fertilizer plants located in the consumption centers of North India.

However, the major policy which has ensured the growth of the fertilizer industry is the thrust on accelerating fertilizer consumption by fixing, on the one hand, low and uniform price for fertilizers, and on the other hand providing the manufacturers adequate compensation through the retention price and subsidy scheme. As expected, fertilizer nutrient demand has gone up from 0.29 million tons in 1960-61 to 13.9 million tons at the end of 1995-96, compared to 12.15MT during 1992-93.

Fisheries

Fish production achieved an all-time high of 4.9 million tons at the end of 1995-96. Programmes that have helped boost production include the National Programme of Developing Fish Seeds, Fish Farmers' Development Agencies and Brackish Water Fish Farmers' Development Agencies. The Central Institute of Fisheries Nautical and Engineering Training trains the necessary manpower. To diversify fishing methods and introduce processed fish products on a semi-commercial scale, an Integrated Fisheries Project has been launched. A National Fisheries Advisory Board has also been established.

Food Processing

A Ministry of Food Processing Industries was established in July 1988 to ensure better utilization of farmers' output by inducting modem technology into the processing of food products, thus augmenting the income of farmers and generating employment opportunities in rural areas. A new seeds policy has been adopted to provide access to high quality seeds and plant material for vegetables, fruit, flowers, oil-seeds and pulses, without in any way compromising quarantine conditions. Initiatives have been taken to encourage private sector investment in the food processing industry.

Agricultural Research

The apex body for education, research and extension education in the field of agriculture is the Indian Council of Agricultural Research (ICAR), established in 1929. India's transformation from a food deficit to a food surplus country is largely due to ICAR's smooth and rapid transfer of farm technology from the laboratory to the land.

ICAR discharges its responsibilities through 43 research institutes, four national research bureaus, 20 national research centers, nine project directorates, 70 all-India coordinated research projects, and 109 Krishi Vigyan Kendras (farm science centers). Besides, the programme of Agricultural Education is coordinated by ICAR with the curricula and other normative guidance given to the 26 Agricultural Universities and four National Research Institutes.

Oilseeds Production

A Technology Mission on Oilseeds was launched in 1986 to increase production of oilseeds in the country and attain self-sufficiency. Pulses were brought under the Technology Mission in 1990. Before the Mission was launched in 1985-86, oilseed production was 10.83 million tons; during 1995-96, it was estimated at 22.42 million tons, which is a record. Soybean, rapeseed and mustard largely contributed the increase in production. Production of pulses has seen many ups and downs, which is expected to be checked under the Mission. The country grows mainly nine oilseeds, with groundnut, rapeseed and mustard accounting for 62 per cent of total production. Lately, soybean and sunflower have shown major growth potential.

Drinking Water

A Technology Mission on Drinking Water and Related Water Management has been constituted to cover the residual problem villages and provide potable water at 40 liter per capita per day, and 70 liters per capita per day in desert areas inclusive of 30 liters for cattle. The Mission is tackling the problem through 55 mini-missions in project districts and countrywide problem oriented sub-missions. A Village Level Operation and Maintenance (VLOM) pump called India Mark-11 has been developed and is being exported to 40 countries. By March 31, 1993, over 79 per cent of the rural and about 85 per cent of the urban population was provided drinking water facilities.

7

Agricultural Production Forms and Capitalist Economy

Another thrust in the neo-Marxian political economy literature, and one that is somewhat contrary, has been the argument that the differentiation of agriculturalists into the capitalist and working classes may be incomplete in the foreseeable future as farm and nonfarm production become integrated into a single system incorporating different organizational forms of production. An early statement of this line of thought was that of the German Social Democrat and Marxist theoretician, Karl Kautsky. Kautsky argued that what was crucial to understanding the evolution of agriculture in advanced industrial societies was not simply the dominant form of ownership of agricultural enterprises, but rather the functions that were served by the emerging organizational formsof agricultural production.

This approach was developed further by Mottura and Pugliesi in an historical analysis of small holdings farmed part-time in southern Italy and the functions of smallholder agriculture in contemporary decentralized economic development. The thrust of their argument was that while most agricultural production took place on farms organized along capitalistic lines, part-time farming served as a backup alternative for the workers in industrial plants located in rural areas. In times of industrial contraction and high unemployment, displaced workers with small farms could turn temporarily to subsistence production until industrial conditions improved, thus forming a reserve labor force.

This integration of agricultural and nonagricultural production spheres has been further elaborated by Bonanno, who examined the role of the state in fostering small farms as one strategy to mediate the interests of the conflicting social classes in the emerging social orders of the advanced societies, particularly Italy and the United States. In this light, various farm programs to deal with problems in the agricultural sector can be seen in part as rooted in the legitimation function served by the continuation of small farms. Small farms are also important in the decentralizing industrial system fostered by state policy. In this system, industrial firms move to rural areas where labor is not unionized and where wages are low because many potential workers have small farms producing inadequate incomes and few alternative opportunities. Work is also increasingly "informalized" in cottage industry, piecework arrangements. In such contexts, small farms serve the function as "keeper of surplus labor" — providing, at the same time, a source of low-cost labor for industry and, in the face of the tenuous employment, a source of security for the members of households with small farms. Wenger and Buck, building on this line of thought but especially on the earlier work of Andre Gunder Frank, take this argument another step by examining how exploitation and superexploitation (extracting more value from workers' labor than permits reproduction of that labor) from members of farm households are necessary and dynamic features of both advanced capitalist societies and developing societies. Production organized according to obligations of kinship ("domestic relations of production") link ("articulate") in various ways with production organized along capitalist lines so that value is transferred from the domestic sphere of production to the capitalist sphere of production.

This takes place in such a way as to make the domestic sphere an "interstitial domestic reserve of labor" that subsidizes the capitalist sphere either directly or indirectly through various mechanisms. For example, unpaid household labor reduces both the wages necessary for workers employed in industry and the prices of agricultural commodities required by farm families. Off-farm income from wage work helps to pay the costs of agricultural production and thus lowers the price of food for other working class families.

FARMERS AS ACTORS IN A CAPITALIST POLITICAL ECONOMY

Whereas the perspectives of Mann-Dickinson and Friedmann on one hand, and of de Janvry, Friedland et al., and others working in the Lenin (and, to a lesser extent, the Kautsky) traditions on the other, are, in a sense, in diametrical opposition, much of the most provocative work in the Marxist tradition in the "new sociology of agriculture" has revolved around formulations that explicitly or implicitly are attempts at synthesis. One of the most noteworthy attempts has been by Mooney.

Mooney's principal contribution has been to cast doubt as to whether the existence of conventional capital-labor relations on farms is an adequate benchmark for gauging the existence of capitalist penetration of agriculture. Following Wright, Mooney has developed a model of agrarian class structure involving "contradictory class locations" such that class positions other than family labor farmer (unity of capital and labor in the farm household), capitalist farmer, and agricultural wage labor are seen to exist. In particular, Mooney sees that there are several "detours" that can be taken by farmers in order to avoid proletarianization (*e.g.*, either being forced from agriculture or becoming hired agricultural laborers).

These detours involve tenancy, contract farming, part-time farming, and debt. In each, whereas there is no capital-labor relation at the point of agricultural production, farmers are exploited by some fraction of nonagricultural capital (in tenancy, by landlords; in contract farming, by agribusiness; in part-time farming, by off farm capitalists; and in debt, by finance capital). Thus, Mooney argues that the exploitation of farm wage workers by agrarian capitalists is only one form that capitalist penetration of agriculture can take. Moreover, Mooney sees that these detours — that is, the alternative ways in which capital acts to "strip simple commodity producers of... surplus value other than through the extension of wage labor" — may be more significant than full-blown capital-labor relations at the point of agricultural production. It is significant, however, that Mooney's explanation of why these contradictory class locations in agriculture tend to emerge has a subjectivist component, based on Weber's distinction between formal and substantive rationality. Mooney sees that many farmers are motivated more by forms of substantive rationality (*e.g.*, the desire for autonomy over their work) than they are by formal capitalist rationality. Accordingly, these farmers tend to be tenacious in holding onto their farms and farm lifestyles and will often tend to take one of the four "detours" to capitalist development in order to remain in agriculture.

Mann and Dickinson, however, replied vigorously to Mooney's neo-Weberianism with two major arguments: first, that so-called contradictory class locations are really based on the same social relations of production grounds that Mann and Dickson use and, second, that Mooney has misconstrued their notion of "obstacles to capitalist development" in agriculture to be immutable barriers, so that Mooney was unable to recognize that his notion of "detours" to capitalist penetration of agriculture is similar to their own views.

Mann and Dickinson thus argue that subjectivism, and the explanatory ambiguities it involves, does not yield insights beyond those afforded by a structural, neo-Marxist theory. They were also critical of Mooney's project of synthesizing Marxian and Weberian approaches on the ground that these approaches are incompatible, except on an ad hoc, eclectic basis. Mooney added further to the debate by arguing that Mann and Dickinson have exaggerated his departure from a neo-Marxist framework-that his perspective retains Marxist insights by incorporating key Marxian categories as ideal-types within a Weberian framework.

He argued that this approach overcomes the incompatibilities of the Marxian and Weberian frameworks — unless, of course, one adopts a mechanistic Marxist or Leninist perspective. His key position was that the presumption of a predictably patterned and necessary differentiation of farmers into the capitalist class and proletariat is erroneous and that subjectively meaningful human action affects the process of differentiation substantially.

With the 1980s farm crisis coming under increasing scrutiny by rural sociologists, Mooney has extended his analytical framework to an historical examination of the state-sponsored farm credit system (Farmers Home Administration [FMHA], Commodity Credit Corporation [CCC], and the Farm Credit System [FCS]) and tax policy incentives for increasing the capitalization of agriculture. Mooney conceptualizes credit and tax policies as means of dealing with the crisis of legitimacy posed by the increasing prevalence of tenant farming produced by the Great Depression. Mooney argues that this credit and policy system created more indebted farmers who ultimately became vulnerable to financial crises in the form of rising real interest rates and declining agricultural commodity prices, a situation such as that which occurred in the 1980s. Financial crises expose to perceptive actors the role of the state in creating the crisis, thus creating another crisis of legitimacy and providing a basis for affected farmers to mobilize against the state.

Pfeffer has taken a considerably different tack than Mooney, though Pfeffer's work has some similarities to that of Mooney in that he stakes out middle ground on the question of whether the predominant feature of agriculture in advanced capitalism is the persistence of the family labor farm on one hand, or its demise into capitalist relations of production on the other.

Pfeffer conducted an historical analysis of three systems of agricultural production in the United States (industrial agriculture in California, family farming in the Great Plains, and the sharecropping system in the South prior to World War II) and demonstrated that recruitment and maintenance of access to a suitable labor force was a crucial factor shaping the disparate organizational structures of these three systems. Only in California, largely because of preexisting land concentration and of state actions to recruit and restrict the residential, educational, and occupational mobility of ethnic minority farm workers, was capitalist agriculture able to set root prior to World War II.

Noting that the U.S. South had a high degree of land concentration during the entirety of the nineteenth century, Pfeffer goes on to explain why southern planters were unable to recruit a wage labor force after the Civil War and why the tying of sharecroppers to plots on the landlord's plantation through the crop-lien system was the only possible route to recruiting the labor necessary to operate a large plantation.

Finally, Pfeffer demonstrates that a combination of labor scarcity, the unwillingness of immigrants to work for wages on farms, and severe declines in the price of wheat in the late nineteenth century led to the establishment and persistence of family farming in the Great Plains. Pfeffer thus suggests that while capitalist agriculture can emerge under particular circumstances, there have been several alternative forms of agricultural organization — sharecropping and family farming, in particular — that have become established due to regional and commodity variations in the ability of farmers to recruit a dependable labor force.

Still a different approach was taken by Whatmore et al., who have developed a theoretically based typology of farms in Britain. Their goal was to understand the effects of external pressures by nonfamily and nonfarm capitals to penetrate agriculture and how these pressures articulated with internal processes on farms. By cross-classifying levels or degrees of subsumption of internal and external relations, they developed a typology with four main categories of farm types, ranging from relatively unsubsumed ("marginal closed unit") to the "subsumed unit." The internal relations variable pertained to ownership and control of farm capital and land, control over management, and the balance of family and hired labor.

The external relations variable pertained to level of dependence on industrial capitals for production inputs and purchase of farm outputs and to the degree of involvement in credit relations, especially with finance capitals. Applying this theoretically based typology to the data from three areas in southern England — an urban fringe area, an agricultural area undergoing commercial and industrial development, and a primarily agricultural area — they found that the process of subsumption of farm production relations, and thus their integration into the "wider circuits of capital," varied over time and space. Whatmore et al. concluded that farm families need to be understood as actors in the process of subsumption rather than its passive victims.

The neo-Marxian theoretical enterprise (and that of the modernization school as well) on the conversion of traditional peasant economic relations to commodity relations under capitalism has also been scrutinized by Vandergeest. Among his major criticisms were that neo-Marxist approaches tend to be too unilinear, too macro-structural, too denying of human agency, too inattentive to the role of the state in the process, and too unconnected with practice. He has argued, from a neo-Weberian position rooted in the work of Pierre Bourdieu, that all categories and theories are in the last analysis historically contingent, ideological, and interpretive social products.

The world is a complex whole which can only be investigated empirically and understood through theory. It cannot be reduced to the simple working out of a model derived through deductive theory — whether that of "simple commodity production" or of "capitalism". There is not a single deductive logic (such as the logic of exchange-value or accumulation) underlying or determining all relations in a capitalist formation, but there are different, historically contingent principles which we can only investigate through empirical research.

There has recently been a tendency in the neo-Marxist and neo-Weberian literatures towards a deemphasis on "deductivist"-functionalist theoretical postures. Subjectivist perspectives such as those of Vandergeest and Mooney have been influential in leading to this reorientation, though this tendency is by no means limited to those persuaded by neo-Weberianism.

The three postures in the Marxist political economy of agricultural tradition — those emphasizing the persistence of petty commodity production, those emphasizing the inevitable differentiation of petty commodity producers into antagonistic social classes, and those seeking synthetic positions — have yielded a continuing, lively debate in the rural sociology literature. Particularly instructive is the debate over the Mann-Dickinson hypothesis. Also of importance is Goodman and Redclift's commentary on the work of Friedmann in which Goodman and Redclift critique Friedmann for taking an overly deductive approach to petty commodity production that ignores the role of historical conjuncture and ideology. Bernstein has made a comparable argument — that generic theories of simple commodity production under advanced capitalism tend to overemphasize its "functional" aspects and downplay contradictions and the diversity of household forms of production.

It should be emphasized that theoretical and empirical treatments of the political economy of North American agriculture have not been made only by rural sociologists. Many of the pioneers in this tradition of scholarship have been nonsociologists or sociologists who have had little or no connection with the Rural Sociological Society or the institutionalized form of rural sociology in U.S. land-grant universities. Nonetheless, rural sociologists have made influential contributions to this literature.

The emerging political economy tradition in the sociology of agriculture has led to a provocative and stimulating literature. It should be mentioned, however, that those working from this perspective are just beginning to establish a distinctive research programme; much of the existing literature, in fact, often tends to involve superimposing a new vocabulary on already-established data, such as Goss et al. set out to do. Establishment of a distinctive research tradition will probably occur slowly since conventional data sources such as censuses and sample surveys are often inapplicable to theoretical issues in the political economy of agriculture. There have, nonetheless, been several research papers in the political economy tradition that have utilized innovative new methodologies to generate research findings that bear directly on specific hypotheses.

Subculture and Agricultural Structure

It is useful to note that the growing interest in subjectivist approaches in the new sociology of agriculture has not been confined to those who work within the neo-Marxist and neoWeberian traditions. Anthropologically-oriented researchers have also contributed to this literature through the examination of the effects of ethnic background on state and local level farm structure. Salamon and associates have done the pathbreaking studies in the anthropological tradition in Illinois communities, though recently this work has been taken up by others.

While their work was not intended to contribute to the "structure versus agency" debate that has raged between Mann and Dickinson on one hand, and Mooney on the other, and actually predated much of this debate, it is clearly relevant to the debate. In particular, the work of Salamon and associates pertains to the role of subjective motivations in farm decision making, especially motivations conditioned by persisting subcultural variations based in differences in ethnic origins of farm families. The key finding of this line of research has been that farm families with German ethnic backgrounds tend to view farming as a way of life and hold a strong value for keeping the family farm intact. These Salamon has labeled "yeoman" farmers. In contrast, farm families with British ethnic backgrounds tend to be more entrepreneurially oriented, viewing farming as a way to make profits and having little attachment to farming or to particular farms. These Yankee farmers are more likely to seek growth in scale and less likely to support their local communities. This research has led to understanding differentials in farm size as a partial function of differences between these subcultural variants. Flora and Stitz found that yeoman farmers in a Kansas county generally expanded less than did Yankee farmers. Foster et al. also found support for the Salamon hypothesis in a study of a sample of Illinois farms that had been in the same families for 100 years or more.

Several other anthropological studies in the tradition of Salamon and colleagues have been reported in Chibnik. Of particular importance is Barlett's article in which she criticizes notions of the demise of the family farm and of the "disappearing middle" from an anthropological perspective.

THE MONOPOLY STAGE OF CAPITALISM SYSTEM

The building of an all-embracing imperialist system was completed at the turn of the century, when the undivided sway of monopoly capital was established in the main capitalist countries, and the territorial division of the world between them was fully completed. The main patterns of the functioning of monopoly capitalism began then to determine the decisive lines of development of the world economy and world politics. The concluding stage of the formation of a world economy, which took around two or three decades, belongs roughly to the same period. The transition irom pro-monopoly capitalism to imperialism was I he result of deep-seated shifts, above all in the productive forces and technical basis of society, under the action of the ever accelerating concentration and centralisation of production and capital in its main centres. The development of monopolies inevitably retarded technical progress, but at tho same time the increasing concentration of their economic and financial power opened up hitherto unprecedented opportunities for broader application of those advances of science and engineering in the world economy that helped them increase their profits.

FORMS AND METHODS OF PRODUCTION

This period was marked by the introduction, especially in heavy industry, of major technical advances that brought into being economically more efficient forms and methods of production. The competitiveness of the monopoly mergers was enhanced and the objective conditions created for a further acceleration of the concentration and integration of production. The technical advances that had a marked effect in changing the structure and accelerating growth of the world capitalist economy included, for example, the invention of new ways of making steel (the Bessemer process in 1855, open hearth process in 1864, and the Thomas, or basic process in 1878). Their introduction by the monopolies of the main capitalist countries made it possible to convert the iron and steel industry into a leading branch of the world economy. Whereas only around 500,000 tons of steel was produced in 1870 in the whole world, production was million tons in 1890, and already 76 million tons in 1913. World production of pig iron rose to 24 million tons in 1890 and to nearly 80 million tons in 1913.

The development of the engineering and chemical industries also had a considerable impact on the advance of many other branches of the economy. The application of technical advances in production that led to the perfecting of prime movers played a revolutionary role in tin's respect. Among them were the invention of the dynamo (1867), the fonr-stroko internal combustion engine (1877), the steam turbine (rnid-80s), and the Diesel engine (end of the 90s). The new types of engine promoted the development of new forms of transport: the motor car (1885), the Diesel locomotive (1891), and the aeroplane (1903). The application of improved means of transport and communication, which had in fact united all areas and countries of the globe at the beginning of the twentieth century, immeasurably accelerated the forming of the world system of division of labour of the imperialist epoch.

In 1913 the length of the world's railways was more than a million kilometres, of which more than 250,000 km had been built in colonial and dependent countries in Asia, Africa, and Latin America. There was also a steep increase in marine freights and ocean transport. The energy base of world capitalism, then based on coal, also grew rapidly. The mining of coal exceeded 1,000 million tons at the beginning of this century. The technical basis of the power industry also broadened considerably. For many decades after Watt's invention of the steam engine capitalism's economic growth was linked with the application of steam, but at the turn of the century the use of electricity was becoming more and more important, which also fostered the introduction of a number of new advances in the field of electrical engineering by the biggest companies: *viz.*, the electric smelting of steel (1877), the electrolytic technique of obtaining non-ferrous metals (end of the 1880s), longdistance transmission of electricity (1891), etc. The world level of production in manufacturing industry rose by 120 per cent between 1880 and 1900, and by 270 per cent by 1913.

DEVELOPMENT OF CAPITAL IN AGRICULTURE

Farm producers have historically been in possession of two powers that stood in the way of the development of capital in agriculture. First, farmers could make choices about the physical process of farm production, including what was grown and how much, and what inputs were to be used. These choices, of course, were always constrained, partly because of local conditions of climate and soil, and partly because of the local nature of markets for farm products. Second, farmers were themselves traditionally potential competitors with the commercial providers of inputs, because they could choose to produce seed, traction power, and fertilizer themselves. The problem for industrial capital, then, has been to wrest control of the choices from the farmers, forcing them into a farming process that uses a package of inputs of maximum value to the producers of those inputs, and tailoring the nature of farm products to match the demands of a few major purchasers of farm outputs who have the power to determine the price paid. Whatever production risks remain are, of course, retained by the farmer. As the farmer loses any power to choose the actual nature and tempo of the production process in which he or she is engaged, while at the same time losing any ability to sell the product in an open market, the farmer becomes a mere operative in a determined chain whose product is alienated from the producer.

That is, the farmer becomes proletarianized. It is of little import that the farmer retains legal title to the land and buildings and so, in some literal sense, is the owner of some of the means of production. There is no alternative economic use for these means. The essence of proletarianization is in the loss of control over one's labour process and the alienation of the product of that labour.

How has this transformation of farming been accomplished? In the first stages, in the century between the invention of reaping machines and the end of the Second World War, innovations in farming directly addressed that problem of the availability, cost, and control of farm labour through mechanization. No farmer could resist the arrival of the tractor, nor could one be home built. After the Second World War, refined and synthetic chemical treatments became the chief purchased inputs in the form of fertilizer, insecticides, and labour-saving herbicides. Again, these purchased inputs could not be resisted because of the large increase in yields and the reduction in labour. Herbicides, in particular, also reduced the requirement for tillage machinery, insecticides reduced the uncertainty of a successful crop, hormone sprays allowed for a close control of ripening time in fruit crops, and antibiotics prevented animal diseases. Once again there could be no competition between these industrial products and self-produced farm inputs.

The analysis of the growing role of capital inputs cannot be made, however, if a central feature of the productive process is lost sight of: the concrete use of all these inputs is to produce living organisms. The steps of mechanization and the use of chemicals were not possible in isolation from the nature of the organisms being produced. Unlike in other sectors of production, in agriculture living organisms are at the nexus of all input streams and are the primary sources of all output transformations. But living organisms are mortal, so their production requires their reproduction. That is, every cycle of farm production begins with seeds or immature animals to which value is added by on-the-farm operations, so seed (or the "seed" animal) is the central input into farming.

The control of the biological nature of these seed organisms is a critical element in the control of the entire process of agricultural production, which puts the provider of this input in a unique position to valorize other inputs. For example, while a dramatic drop in the price of nitrogen fertilizer at the end of the Second World War made it economically possible for farmers to use this input in large quantifies, for this input to be useful it was necessary to breed plants, corn hybrids in particular, that could, in fact, turn a massive nitrogen application into crop yield.

The successful mechanization of tomato harvesting was only possible by a close cooperation of the machine designers with plant breeders. The breeders completely remade the biology of the tomato plant, turning a loosely branched plant that flowers and sets easily bruised fruit continuously over the growing season into a short, stout, Christmas tree-like plant whose tough fruits all ripen at about the same time.

The consequence of the central position of seed input in the production process is that seed companies are potentially in an extraordinarily powerful position to appropriate a large fraction of the surplus in agriculture. There is a barrier to this realization however. The seed of a desirable variety, when planted by the farmer, produces plants that themselves produce yet more seed of the variety.

Thus the seed company has provided the farmer with a free good, the genetic information contained in the seed, that is reproduced by the farmer over and over again in the very act of farming. Some way must be found to prevent the farmer from reproducing the seed for next year's crop. The historical answer to this problem was the development of the inbred/hybrid method of breeding, using hybrid crosses between inbred lines, which makes it possible to sell seed that will produce hybrid plants, but which themselves do not reproduce hybrids. Because the second generation would not be true hybrids and thus would lose yield and be more variable, the farmer must go back to the seed company every year to buy new seed. As a result of the immense profits made by seed companies selling hybrid maize seed, the method was spread into other organisms such as tomatoes and chickens. Moreover major commercial hybrid seed and chicken breeders like Dekalb, Funk, and Northrup-King were at one time acquired by pharmaceutical and chemical companies like Ciba-Geigy, Monsanto, and Dow, although subsequently there have been divestments and realignments. Only the largest hybrid seed company, Pioneer Hy-Bred, remained obdurately independent until, in 1997, 20 percent of its equity and two seats on its board were purchased by DuPont.

Generally, the ability of commercial seed companies to control seed inputs by the inbred/hybrid method was severely limited. First, the method cannot be made economically workable in many important crops like soybeans and wheat, or in large animals. Second, while the inbred/hybrid method was successful for general yield increase, large numbers of important specific characters such as resistance to particular diseases, or resistance to herbicides, or increases in oil content in oil seeds, do not show hybrid vigor, but must be introduced by other breeding methods. Third, there are characteristics that would be desirable to introduce into an agronomically important species, but which are present in other organisms that cannot breed with the species under cultivation. The most famous example was the desire to make corn plants able to fix nitrogen from the atmosphere, as legumes are able to do, by making their roots hospitable to nitrogen-fixing bacteria. While this would reduce the market for nitrogenous fertilizers, it would place the provision of nitrogen in the hands of seed companies!

The limitations on what changes could be made to agronomic species that would be profitable to seed companies and their chemical company partners or owners meant that the penetration of capital into agriculture had reached its apparent limits in the 1970s. The introduction of major new forms of mechanization into farm production had come to an end, partly because of the dramatic change in the cost of fuel and partly because a steady supply of immigrant labour that could be deported smiled progress in agricultural labour organizing. The growing public consciousness of the polluting effects of fertilizers and pesticides and the development of OSHA regulations to protect farm workers against the deleterious effects of insecticide and herbicide sprays discouraged radical changes in the uses of chemicals or even continued growth in the use of older materials. In addition, these fertilizers and pesticides were being used at very high rates, probably higher than could be economically justified by the farmer.

There was, for example, no growth in fertilizers after 1975 or in synthetic pesticide application rates beginning in about 1980. Any further possibility for input providers and output purchasers to increase their appropriation of the surplus in agriculture depended on (1) making some radical changes in the biology of agronomic species, and (2) guaranteeing that such changed biological systems would remain within their ownership and control. Moreover, that appropriation could be greatly increased by a greater consolidation of both input and post-farm production sectors (purchasing, processing, and distribution), to provide near monopoly control. Enter biotechnology.

SOCIALIST MODE OF PRODUCTION

After World War I there was a comparatively rapid restoration and further expansion of production, and of the world economic links broken by the war, within contracted boundaries. But this growth took place on a background of further deepening of imperialism's basic contradictions, a sapping of its colonial foundations, and an intensification of the unevenness in the development of separate regions and countries. At the same time the socialist mode of production was more and more clearly becoming a constituent part of the global economy of mankind, and was graphically demonstrating its vitality. The international standing of the Soviet Union was strengthened, and its external economic 51 relations extended, through intergovernmental agreements with several capitalist conn tries, certain former colonies and semi-colonies. The progress made in building socialism in the USSR, and the consolidation of its position on the international arena, where the foundations of a new world social system were being laid, became vital factors in the subsequent aggravation of the general crisis of capitalism.

Growth of capitalism's irreconcilable contradictions continued to play the determinant role in this process. Particularly convincing evidence of that was the world cyclical overproduction crisis of 1929–33, the greatest in scope, duration, and disruptive consequences in capitalism's history, which simultaneously swept the industrially developed countries and their agrarian, raw material periphery. The crisis, which threw the economy of the capitalist world several years back, had a very negative effect as well on the whole system of imperialism's global economic connections. By the end of the 30s the volume of international trade, and also of capitalist industrial production was below the level of 1913. Right up to the beginning of World War the economies of most countries of the capitalist world had still not been able to recover fully from that crisis.

At the same time the colonial foundations of the capitalist economy had-been markedly weakened. The successful anti-imperialist struggle of the peoples of the USSR and their achievements on the road of radical socio-economic reforms, stimulated an upsurge of the national liberation movement in many colonies and dependent countries, some of which had managed after World War 1 to win or retain their national independence. A crisis of the all-embracing colonial system began. The colonial powers, it is true, were then still able to avert the break-up of their empires. Without ceasing their struggle to redivide spheres of domination, they endeavoured as before to deprive the countries economically dependent on them of national sovereignty; and to that end imperialism repeatedly unleashed predatory wars (*e.g.* Italy's aggression against Ethiopia, and Japan's against China).

Characteristically, however, not a single imperialist power could succeed, even in the initial stage of the general crisis of capitalism, in turning any economically backward country into a new colony. While retaining a decisive position in the world economy, linance capital was forced to take into account in its expansionist policy not only the 'strength' and 'capital' of its imperialist rivals but also of the growing strength of resistance of the peoples of the countries exploited by them, who received every possible support from the iirst socialist country. World War II, which broke out as a result of the sharp aggravation of all capitalism's contradictions at the end of the 30s, and of the striving of its most aggressive circles to strike a crushing blow against socialism and the national liberation movement, in the final count led to a further deepening of the crisis of the imperialist system, the boundaries of which contracted even further. The socialist mode of production became predominant in a number of countries in Europe and Asia, and a new phase opened in the economic competition of socialism and capitalism.

The antagonism and interaction of the two opposing social systems became a determining feature of today's world economy in the postwar years; in that connection it has become necessary, when studying the structural shifts and fundamental features of the postwar development of capitalism, to take into account at the same time the global tendencies of the growth of production, the more so that the long-term consequences of the scientific and technical revolution that has embraced the leading spheres of the productive activity of human society as a whole are beginning to have a mounting effect on both socialist and capitalist countries, and also on many developing ones. Analysis of the facts characterising these tendencies indicates that in the period under review mankind's productive forces have attained a height hitherto unprecedented. The international exchange of scientific and technical knowledge and production know-how has reached a fundamentally new level. The infrastructure of world economic relations has also developed rapidly under the impact of the scientific and technical revolution on the basis of a national and international division of labour. On the whole, the volume of the aggregate product of all the countries of the world was more than four times as big at the end of the 70s as at the beginning of the 50s. There was also a marked increase in the main productive forces, *viz.*, manpower (the 53 number of the world's inhabitants rose by almost 75 per cen! in the same period).

The list of that kind of quantitative indicators could, of course, be considerably extended, but the figures cited give a graphic idea of the sizable shifts that have occurred in humanity's productive forces in the contemporary historical situation of competition of the two main modes of production and the corresponding two world socio-economic systems, namely the socialist and the capitalist. As Karl Marx remarked in a letter to P. V. Annenkov (28 December 1846), tho productive forces are therefore the result of practically applied human energy; but this energy is itself conditioned by the circumstances in which men find themselves. Which of the opposing world social systems has the advantage in creating conditions for growth of the productive forces can be judged from the following facts. On the whole, in the three decades of the 50s through to 80s, the rates of growth of national income and industrial output in the socialist countries belonging to the Council for Mutual Economic Assistance (CMEA) were three time as high as *in* all the capitalistically developed countries taken together.

The countries belonging to CMEA are the most dynamically developing economic community in the world. At the end of the 70s these countries, which have 10.2 per cent of tho world population, produced around one-third of the world's industrial output. In 1968–79 alone the volume of the gross social product in the European members of CMEA increased more than fourfold, while that of all the developed capitalist countries increased by only 140 per cent. A natural consequence of the development of this tendency has been a considerable raising of the role of socialism in the world economy. At the beginning of 1980 the population of 54 countries not yet broken free from direct, colonial dependence was not more than 0.1 per cent of all mankind.

The turn of the decade to the 80s was marked by further growth of the economic might of the countries of world socialism. As the Central Committee's report to the 26th Congress of the Communist Party of the Soviet Union noted: The past few years have not, been among the most favourable for the national economies of some socialist states. Still, in the past ten years the economic growth rates of the CMEA countries have boen twice those of the developed capitalist countries. Tho CMEA members continued to bo the most dynamically developing group of countries in the world.

At the same time the break-up of the colonial system was not limited to emancipation of the colonies, but also included semi-colonial and dependent countries. Throughout the long period of'the imperialist powers' unlimited sway in the world economy the dominant tendency was to convert the latter into rightless colonies; in the postwar years another trend began to predominate, that of converting them into politically independent national states. Imperialism could not suspend the development of this tendency as well. In recent decades the overwhelming majority of semi-colonies and dependent countries, having thrown off foreign and proimperialist regimes, have also in fact broken free of imperialism's system politically, and are no longer its reserve, as they used to be. With the all-round support of the socialist countries and international working-class movement, they are intensifying the anti-imperialist struggle for economic independence.

As a result there has been a marked quickening of the growth rates of their economies. The total production of goods and services in the developing countries of Asia, Africa, and Latin America rose more than 320 per cent in the 50s to 80s, industrial output rising by 620 per cent and farm production by nearly 130 per cent.

Their position in the world capitalist economy, however, remains extremely difficult and unequal. In spite of the fact that they have three-quarters of the population of the non-socialist world, they produced less than one-fifth of 55 that world's gross output at the beginning of the 80s. The immense gap in levels of development between them and the group of imperialist powers historically built up under capitalism has continued to widen. In the first postwar years, for example, this gap, in terms of per capita GDP, was characterised by a ratio of 1: 10, and at the beginning of the 80s by a ratio of more than 1: 13. In the same period there was a marked reduction of the weight of the former colonial world in the turnover of the world capitalist market. The capitalist economic system is more and more clearly demonstrating its incapacity to deal with the pressing socioeconomic problems of emancipated countries. The logic of history is convincing their peoples of the lack of perspective of development along the capitalist road.

It is not fortuitous that many of the new national states, including some that are still largely enmeshed in imperialism's world economic relations, are rejecting the capitalist road of development in their constitutions, under pressure of the broad public, and are proclaiming a line of building a socialist society in the long term. Leonid Brezhnev, characterising the significance of the countries liberated from colonial subjection that have taken the road of revolutionary, democratic transformations, and of countries with a socialist orientation in today's world, said that at the 26th Congress of the CPSU:

Development along the progressive road is not, of course, the same from country to country, and proceeds in difficult conditions. But tho main lines are *similar*. These include gradual elimination of the positions of imperialist monopoly, of the local 56 big bourgeoisie and the feudal elements, and restriction of foreign capital. They include the securing by the people's state of commanding heights in the economy and transition to planned development of the productive forces, and encouragement of the cooperative movement in the countryside. They include enhancing the role of the working masses in social life, and gradually reinforcing the state apparatus with national personnel faithful to the people. They include anti-imperialist foreign policy. Revolutionary parties expressing the interests of the broad mass of the working people are growing stronger there.

MONOPOLY CAPITALISM

Monopoly capitalism is really losing its old opportunity to decide the fate of nations even within the shrunken framework of the world capitalist economy. The main point in the people's national liberation struggle at the present stage of the general crisis of capitalism"is that it has been converted in practice into a fight against exploiter relations in general, both feudal and capitalist, in many countries. As Leonid Brezhnev has put it:

It is of exceptional importance that many of the countries that have achieved liberation have rejected the capitalist road of development and adopted a socialist orientation, setting themselves the goal of building a society free from exploitation. And that, without doubt, is a very strong blow to capitalism's positions as a whole as a world social system. A substantial socio-economic consequence of the collapse of the colonial system is the relative lowering of the role of expatriate capital in the economies of emancipated countries. Before independence almost all their industrial production was in the hands of foreign monopolies and was part of the economy of the imperialist system. Today broad measures have been taken in'many of them both to nationalise expatriate enterprises and establish public control over their activity, and to found their own national industry. There are few summarised, reliable facts in the international statistics on what proportion of industry remains under the direct control of expatriate monopolies, but varrious indirect indicators suggest that a considerable^part of the industrial potential of the developing countries, yielding more than 4 per cent "of gross world industrial output, has already passed *out* of such control.

Because of the effect of these tendencies, a further weakening of the imperialist powers' positions in the world economy has become typical of the period of the general crisis of capitalism, which finds expression above all in the obvious undermining of their former hegemony in world industrial production. At the end of the 30s the monopolies of the capitalist countries (including colonies and dependent countries) in practice controlled around nine-tenths of total world industrial production; at the beginning of the 80s only a little more than half of it was under their control. The facts adduced here primarily bring out one of the decisive trends in the crisis of the imperialist economy, when the basic features and main peculiarities of contemporary world development are beginning to be determined by the working class and its offspring, the world socialist system, and no longer by the monopoly capitalists. The steady strengthening of the economic might of the socialist countries, the advance of the national economies of the new sovereign states, and the rapid development of mutually advantageous international economic relations between them are leading in turn to a marked limitation of the imperialist monopolies' diktat in world economic relations, including those between developing and developed capitalist countries on the world market.

Even the most persistent and subtle defenders and propagandists of the capitalist system are unable to deny the tendencies, which are finding wide reflection in international statistics, and which express the main trends in the deepening of the contemporary crisis of the world capitalist economy. An ambivalent approach to these processes is therefore typical of capitalist politicians, economists, and sociologists, who are engaged in apologetics for imperialism. On the one hand, they prefer in every way to evade comparison of the long-term results of the development of the two world economic systems in their interconnections, and strive to avoid a comprehensive socio-economic analysis of these results. While the work of some leading capitalist scholars 58 undertakes to examine the dynamics of l4io economic growth of the opposing social systems today, it usually mentions the advances of socialism in passing as something irrelevant for an evaluation of the present-day level and the ensuing outlook for the global development of mankind's productive forces. On the other hand, immense attention is paid in Western capitalist literature to another aspect of this problem, namely, the factors that continue to promote growth of production within the capitalist economy. These factors are usually considered out of context of the processes showing overtaking growth rates in the economy of the world socialist system.

The ideologists of the capitalist system, glossing over its class and other antagonisms in every way, are relying on the development trends of social development in its industrial centres under the impact of the developing scientific and technical revolution. What lack of prospects can there he talk of, they ask, and even more what hopeless crisis of this social system, if capitalism's productive forces still continue to grow, albeit unstably and with sharp fluctuations.

The increase in the production capacity of the postwar world capitalist economy is thus advanced as the 'weightiest' argument to substantiate the 'historical stability' of capitalism and consequently, as well, the flimsiness of the Marxist-Leninist analysis of the irreconcilable contradictions of its development. In fact the Marxist-Leninist theory disclosed another causal connection between the growth and internationalisation of social production under capitalism and its historical destinies, than that which its capitalist critics try to ascribe to it. The objective tendencies in the development of the capitalist mode of production are not leading to a stopping of growth of the productive forces, or to their stagnation, but on the contrary are leading to their accelerated advance compared with preceding formations. In the *Manifesto of the Communist Party* Marx and Engels wrote:

The bourgeoisie cannot exist without constantly revolutionising the instruments of production, and thereby the relations of production, and with them the whole relations of society—Constant revolutionising of production, uninterrupted disturbance of all social conditions, everlasting uncertainty and agitation distinguish tho bourgeois epoch i'rom all earlier ones.

But does all that mean that, as the productive forces of capitalist society develop, its economic basis is perpetuated and becomes ever firmer and more unshakeable? The founders of Marxism-Leninism, in bringing out the inner contradictory dialectic of that development, demonstrated that it in no way contributes in the end to the consolidation of capitalism. It relentlessly undermines the foundations of this system, and forms the conditions within it necessary, for victory of a more advanced social system. By accelerating growth of the productive forces in its own class interests and internationalising them in every way, the capitalist class itself prepares the material and social conditions for destruction of the social formation created by it.

The transition to monopoly capitalism, while heightening the unevenness and spasmodic character of its development, was marked by a further growth of capitalist production. In describing imperialism as the highest and last stage of capitalism, in which tendencies towards decay and growth of society's productive forces predominated simultaneously, Lenin stressed that:

It would bo a mistake to believe that this tendency to decay precludes the rapid growth of capitalism. It does not. In the epoch of imperialism, certain branches of industry, certain strata of tho bourgeoisie and certain countries betray, to a greater or lessor degree, now one and now another of these tendencies. On tho whole, capitalism is growing far more rapidly than before; but this growth is not only becoming more uneven in general, its unovenness also manifests itself, in particular, in tho decay of tho countries which are richest in capital.

The very substantial shifts in capitalism's material basis in its imperialist stage graphically confirm the fundamental character of these tenets of Marxist-Leninist theory, which help us better to understand the action of the long-term trends in the world economy that predetermine the features and course of its development in today's transitional period. As the general crisis of capitalism deepens, the world capitalist economy more and more sharply experiences the effect of the socialist and national liberation revolutions, the mounting militarisation of the economy, and of numerous military shocks and upheavals. The development of 60 production in all the countries of this economy has been repeatedly interrupted or held back by economic crises, but the objectively operating long-term tendency towards growth of society's productive forces has still continued to make itself felt in the stage of imperialism as well. The economies of capitalist countries began to develop at particularly high rates after World War II, as witness the following facts. Fundamentally new branches of industry, transport, and communications were developed then under the impact of the scientific and technical revolution. Internationalisation of social production was noticeably broadened and deepened; the physical volume of international trade on the world capitalist market rose nearly eightfold.

The number of working people, and their vanguard the working class, increased on an unprecedented scale in the postwar decades. When Marx and Engels disclosed the very great revolutionary potential of the proletariat in the *Manifesto of the Communist Party*, there were only around nine million workers in the whole world. At the beginning of the 1980s their numbers had increased nearly 50-fold. The working class has become the leader of the broad masses' light for a revolutionary restructuring of society. And only in that struggle could the new social system arise that is going to replace capitalism. Capitalism, Lenin said, creates its own grave-digger, itself creates the elements of a new system, yet, at the same time, without a 'leap' these individual elements change nothing in ths general state of affairs and do not affect the rule of capital. The patterns of the general crisis are not introduced into the world capitalist system from outside but are the natural result of its inner development and stem from the antagonistic nature of the capitalist mode of production itself, based on the exploitation of man by man and of some countries by others. The decisive consequence of the structural shifts under review was not a weakening but an unprecedented sharpening of the basic contradiction of capitalism, *i.e.* the contradiction between the constantly increasing social character of production and the capitalist, exploiter character of the appropriation of social labour. It is on that basis that all the other contradictions of modern capitalism and its world economy continue to develop. A thorough study of the capitalist economy of his time enabled Karl Marx to conclude that the historical development of the antagonisms, immanent in a given form of production, is the only way in which that form of production can be dissolved and a new form established.

Lenin repeatedly stressed the paramount importance of extending this conclusion of Marx's to analysis of the monopoly stage of capitalism, and pointed out, already at the beginning of the contemporary transitional epoch that: It would be impossible to put an end to the rule of capitalism if the whole course of economic development in the capitalist countries did not lead up to it. Capitalist economics lias repeatedly tried in this connection to give battle to Marxist-Leninist political economy on the problems of world economic crises, in which, as Engels put it, 'the contradiction between socialised production and capitalist appropriation ends in a violent explosion'.

CAPITALIST MODE OF PRODUCTION

The characteristics of the most important stages of the long forming and evolution of this system taken together with the genesis of the capitalist mode of production, the world market, and the aims and methods of colonial exploitation of some countries by others, constitute the main substance of Part I of this book, in which special attention is paid to defining the object of study and the internal contradictions and long-term trends of its development that shape the basic causes of the rise and inevitable deepening of the crisis of the world capitalist economy today. The data adduced will provide the necessary theoretical foundation for bringing ont the themes of the succeeding chapters. The general results and decisive trends of the development of the postwar world capitalist economy, all the ecanomic factors and social consequences of the long-term shifts in the structure and dynamics of both the social production and the consumption of all the countries of this economy taken together.

Then follows a detailed comparison of two groups of countries (*viz.*, capitalist and developing) in order to bring out the specific character of their postwar development, including the impact of the scientific and technical revolution and break-up of the colonial system on their economies. The trends in the growth of population and manpower have been analysed from the same standpoint, and also the productivity of social labour in the main geographical regions of the non-socialist world. All this has enabled us to attempt a more circumstantial and detailed estimate of the changes that have come to light in the regional structure of the distribution of the leading sectors of social production in both the industrial centres and the agrarian and primary producer periphery of capitalism. The problem of the cyclic movement of the postwar world capitalist economy has a place of fundamental importance in the book. Analysis of it helps clarify many features of the regular intensification in recent decades of the uneven economic growth and structural instability of modern capitalism's international business relations.

The main lines of the intertwining development of the industrial and agrarian and raw material base of the postwar world capitalist economy are analysed in Part III. Study of the changes in the manufacturing industries in capitalist and developing countries that are most significant as regards their socio-economic consequences occupy the foreground. Then the long-term shifts in the dynamics, volume and structure of the production of raw materials and in the movement of prices of industrial items, farm produce and raw materials on the world capitalist market are examined. On that basis a number of objectively operating trends are brought out, in which the essential features of the present stage of the crisis of the imperialist system of international division of labour are being more and more clearly manifested within the framework of the world economic relations of recent years. It would not be legitimate to appraise all the trends studied in isolation from the decisive patterns of the world 9 revolutionary process of modern times. As Leonid Brezhnev stressed at the 25th Congress of the Communist Party of the Soviet Union:

This is an epoch of radical social change. Socialism's positions arc expanding and growing stronger. The victories of the national liberation movement are opening up now horizons for countries that have won independence. The class struggle of the working people against monopoly oppression, against the exploiting order, is gaining in intensity. The scale of the revolutionary-democratic, anti-imperialist movement is steadily growing. Taken as a whole, this signifies development of the world revolutionary process.

This book does not claim, of course, to be an exhaustive analysis of the problems posed; many of them certainly need further study, tidying up, and concretisation. Because of the inadequacy or scrappiness of reliable statistics and general economic data, separate propositions of the themes are reviewed in outline or are advanced tentatively. When the conclusions are based on published work or well-known data the exposition has been compressed to a resume, and references are given to the sources containing the detailed argumentation.

The Russian original appeared in 1978. In the present English edition the author has tried to continue analysis of the problems and trends studied, employing the latest data available to him. On that basis certain amendments have been made, and the conclusions and estimates of the probable outlook for the development of the world capitalist economy have been elaborated and broadened.

CAPITALIST MODE OF PRODUCTION AND THE WORLD MARKET

The continuous interaction of the development of production and the world market had already come out in the early stages of the rise of the capitalist social formation, and the determinant place in this interaction, moreover, was taken by capitalist industry. As a result of his thorough investigation of extensive historical data, Marx came to the conclusion that when in the 16th, and partially still in the 17th century the sudden expansion of commerce and emergence of a new world market overwhelmingly contributed to the fall of the old mode of production and the rise of capitalist production, this was accomplished conversely on the basis of the already existing capitalist mode of production. The world market itself forms the basis for this mode of production. On the other hand, the immanent necessity of this mode of production to produce on an ever-enlarged scale tends to extend the world market continually, so that it is not commerce in this case which revolutionises industry, but industry which constantly revolutionises commerce. Marxist political economy consequently looks upon the rise and development of the world capitalist market as a constituent of a broader process, *i.e.* the gradual conversion of capitalism into a universal socio-economic system. This methodological approach has enabled it to disclose the unequal character of the division of labour between countries that is inherent in this system, and the exploiter nature of commodity exchange and of capitalist international economic relations in general on the world market, based as they are on private property in the instruments and means of production.

The struggle for foreign markets has always been an urgent economic need for capitalism, since a striving to extend its sphere of influence without limit is inherently characteristic of it, in contrast to previously dominant modes of production. Because of the operation of objective economic laws, capitalist enterprise inevitably outgrows the boundaries of the community, region, or country, the economic exclusiveness and isolation of countries being broken down by commodity circulation itself.

As a result, capitalist production becomes inconceivable without foreign trade and world economic relations at a certain stage of its development, which in turn reflects growth of the social division of labour on an international scale. Marx stressed that capitalist production rests on the *value* or the transformation of the labour embodied in the product into social labour. But (his is only [possible] on the basis of foreign trade and of the world market. This is at, once the precondition and the result of capitalist production.

He expressed the same idea even more succinctly in the formula 'capitalist production does not exist at all without foreign commerce'.

All that in no way signifies, of course, that any capitalist enterprise indispensably works for an external market and sends what it produces beyond the local or national economy. At the same time it constantly experiences the effect of the already existing system of international business relations and is drawn into their orbit, if not directly, then indirectly. The capitalist, even the one who is wholly oriented on the home market, is forced to compare his own costs of production with world prices as well as with the market prices in his country. An influence of the same order has to be allowed for when an all-round appraisal is being made of the position of labour power in capitalist countries, including the spheres where its value depends wholly or predominantly on domestic factors of production.

Capitalism's international economic interconnections arose later, historically, than its domestic ones, the latter being original and basic. Their level and trend depended to a decisive degree on how far each nation developed its own 21 productive forces, division of labour, and internal dealings. That, however, does not eliminate the undoubted fact that extension of both the home and foreign market under capitalism represents different aspects of a single process; the boundaries between them are very mobile and arbitrary.

As the world market develops the problem of the limits of the home market of each country taken separately proves to lie very tightly bound up with that of extension of its external economic relations. Hence one of the essential principles of Marxian researcli into the development of capitalism in scope and depth is that, when analysing international relations of production (including the capitalist class's colonial policy), it is not important where the boundary between the home and foreign market is drawn; what is important is that capitalism cannot exist and develop without constantly expanding the sphere of its domination, without colonising new countries and drawing old non-capitalist countries into the whirlpool of world economy.

The rise of capitalism's international exploiter relations is inextricably linked with the forming of a system of world economy appropriate to them.

The watershed on this point between Marxian and capitalist science is clear and distinct. Although the creation of this system was completed on the whole under imperialism, it by no means follows that its present-day history can be separated from the preceding history of the development of the capitalist mode of production and that imperialist exploitation of some countries by others begins only with capitalism's transition from free enterprise to the domination of monopolies. Many of the features of the international division of labour typical of the monopoly stage had already arisen and begun to take on a stable, long-term character during the transition of capitalist society to the machine stage of production. They were not brought about by chance, transitional factors but wore caused by deep-seated changes in both the home and external conditions of capitalist reproduction.

The analysis of international economic inter-relationships made in their day by the founders of Marxist political economy indicated that it was large-scale machine industry that prepared the ground for the subsequent formation of a world economy. Machine production, as Marx and Engels pithily put it, produced world history for the first time, insofar as it made all civilised nations and every individual member of them dependent for the satisfaction of their wants on the whole world, thus destroying the former natural exclusiveness of separate nations.

It is through the international division of labour on the world market that the tendency to bring countries together economically, constitutionally inherent in capitalism, is manifested. The kernel of this pattern, however, discovered by Marx, is that this tendency, while progressive in itself, led at the same time, because of capitalist methods and the forms of economic association of the different countries, to consolidation of the division of the world into industrially developed countries, and economically backward agrarian and primary commodity producing countries exploited by them. The antagonism between these two main groups of countries, which are opposite poles of the world capitalist economy, has become one of its main contradictions.

This contradiction has become specially acute in the stage of imperialism, and a very important factor in subsequent origin and development of the crisis of the colonial structure of the capitalist economy, a crisis that is irreversible for the monopoly capitalist class. The break-up of the colonial system after World War II, and the marked deepening of the internal contradictions of world capitalism, which are undermining the historically formed foundations of its system of unequal international economic relations, have put on today's agenda, with all urgency, the issue of eliminating the exploiter system of division of labour that completely dominated the world economy in the past.

It would be wrong to treat this task (which corresponds to the vital interests of all nations without exception) as the need to liquidate the progress already made in the internationalisation of social production. The anti-imperialist struggle for genuine national independence, just like the transition of countries to a higher socio-economic system, does not eliminate the objective need to develop the social (including the international) division of labour, and leads to a change in and perfecting of the forms in which this essentially progressive process is manifested. It is self-evident, Marx wrote to Ludvig Kugelman on 11 July 1868, that this necessity of the distribution of social labour in definite proportions cannot possibly be done away with by a *particular jiirin* of social production but can only change the mode of *its appearance.*

The underlying principles of a Marxist-Leninist analysis of the ways and causes of capital's internationalisation of social production have not lost their theoretical value over the years. The topicality of many of them, moreover, is growing, especially in today's situation of the coexistence and opposition of two world social systems and of the natural extension of economic and technical co-operation on the basis of mutual advantage and equal international division of labour.

In this situation capitalist ideologists are trying in every way to prove the 'erroneousness', in particular, or at least 'obsoleteness' of the Marxist-Leninist political economy's analysis. Above all they endeavour to refute its conclusions about the irreversibility of tendencies towards a steady deepening of capitalism's internal contradictions. But the whole long history of capitalist economics, and of its mistakes, its constant wavering between one conception and another, its fallacies and failures, shows that theories that consider the international market relations of capitalism to be legitimate and that are abstracted from a socio-economic analysis of the production sphere, at best contain superficial analogues of the real trends of development. In the main they bring out the consequences but not the causal connections of this development.

At the same time it would clearly he an oversimplification of things to consider that the theoretical findings of capitalist political economy in the sphere of the world economy's market relations are in general isolated from real analysis of the reproduction processes. In a numher of cases they more or less adequately reflect important features of the backlash of these relations in the production sphere. And this effect determines many of the essential features of the capitalist economy at the various stages of its development. In the last instance (Engols wrote to Conrad Schmidt on 27 October 1890) production is the decisive factor. But as soon as trade in products becomes independent of production proper, it has a movement of its own, which, although by and large governed by that of production, nevertheless in particulars and within this general dependence again follows laws of its own inherent in the nature of this new' factor; this movement has phases of its own and in its turn reacts on the movement of production.

All this calls for close attention to research both into the ways and determinant tendencies in the genesis of the world market and into its place in the process of extended capitalist reproduction. Not only is such research of educational value historically; it is also very necessary for an analysis of current changes in the structure of the world capitalist economy. A systems analysis of its inner and external interconnections in today's international situation cannot be made sufficiently full without detailed allowance for tiie general patterns and long-term trends of development of these interconnections.

Lenin more than once stressed the need to study broad socio-economic phenomena in historical retrospect, since the most important thing if one is to approach this question scientifically is not to forget the underlying historical connection, to examine every question from the standpoint of how the given phenomenon arose in history and what were the principal stages in its development, and, from the standpoint of its development to examine what it has become today.

There were relatively broad trade and financial relations between separate countries and peoples, of course, in the socio-economic formations that preceded capitalism, but it was capitalism that first formed a world market which played a progressive, revolutionary role in the subsequent increase of mankind's productive forces and growth of its social production. Marx, attributing paramount importance to that, wrote that 'the specific task of bourgeois society is the establishment of a world market'. It was with the formation of a world market that the level of development of capitalism began to be marked by the degree of growth both of domestic and of international market relations.

There was a qualitatively new leap in the shaping of these relations associated with the vast increase in trade and colonial expansion of the young European capitalist class. This expansion became a decisive factor in the primitive accumulation of capital. The world market thus created important material preconditions for victory of the capitalist modeof production over the feudal.

At that time the merchant capital of European powers undoubtedly prevailed in international economic relations, but its dominance on the world market began to decline as the capitalist mode of production was consolidated in the countries that were the basis of its existence. The subordinating of merchant capital to industrial, natural to the new mode of production, was also manifested then in the realm of international trade. As a result the material conditions were laid for an ever increasing intensification of the action of the mechanism of the economic exploitation of all the countries of the world by a handful of 'advanced' countries. And the wider the gap in levels of labour productivity becomes between this handful of advanced countries and the economically backward ones, the greater is the exploitation suffered by the latter on the world market. This pattern was expressed with maximum clarity in the following formula of Marx:

The favoured country recovers more labour in exchange for less labour, although this difference, this excess is pocketed, as in any exchange between labour and capital, by a certain class.

In that way, in particular, the main antagonistic contradiction of the capitalist mode of production, *i.e.* that between its constantly increasing social character and the private capitalist forms of appropriating the results of social labour, thus found expression within the world market.

The putting of capitalism onto the rails of large-scale machine industry during the industrial revolution, which was completed in the second half of the nineteenth century, meant the beginning of a new, higher stage in the development of world economic relations. Large-scale industry in fact shaped the world market of the epoch of mature capitalism. And its dependence on external economic links and the international division of labour in turn also grew immensely. As Karl Marx said in his day: Large-scale industry, detached from the national soil, depends entirely on the world market, on international exchange, on an international division of labour.

SITUATION THE ANTAGONISTIC CONTRADICTIONS OF THE WORLD MARKET STEMMING

In this situation the antagonistic contradictions of the world market stemming from the basic economic contradiction of capitalist society, have become an ever more serious obstacle to a further rise of social production. One of the most convincing bits of evidence of that were the world economic crises that were converted into a constitutionally inalienable feature of the whole economic system of capitalism. Years of comparatively rapid industrial boom began to alternate with regular consistency with periods of stagnation and direct destruction of already created productive forces. This cyclicity, determined by the objective laws of capitalist production, was intensified in turn by industry's mounting dependence on the foreign market. It was their comprehensive study both of the internal, home factors of the cyclic development of capitalist business and of the international ones that enabled the authors of classical Marxist political economy to draw the conclusion, confirmed by the whole subsequent course of social development, that economic crises would remain, while capitalism existed, one of the incurable ills determining its historically transient character.

For more than a century and a half now, since 1825, capitalism has periodically suffered from cyclic crises. All the attempts of capitalist economists to work up prescriptions to cure this illness without eliminating its basis, namely its exploiter relations of production, have unfailingly proved unsound and bankrupt. The capitalists as a class have demonstrated more and more clearly with the course of time their incapacity to control the productive forces of society rationally not only within the context of the separate national economies but also and especially on the scale of the world economy.

The world cyclic crisis of 1974–75, the greatest since the war, which developed under an unprecedented sharpening of several of modern capitalism's most important world problems, *viz.* the monetary, energy, raw material, and food crises, was convincing evidence of that indisputable fact in our day. It swept almost all the main centres of capitalism simultaneously and in turn interrupted their economic development. During the crisis the monopoly capitalists as usual exerted immense efforts to shift its most serious consequences onto the working masses of their own countries and the peoples of the agrarian and primary producer periphery.

As capital internationalises society's productive powers the connection between the reproductive processes of the individual countries (including those that are the epicentres of world crises under the impact of the economy's cyclic movement) is becoming closer and closer. Therefore, when we are determining the objective patterns of capitalism's development, the question of the interaction of the national cycles within the context of a single world cycle (study of which was begun by Karl Marx) is posed in all its acuteness. We shall dwell in detail on the features of the cyclic development of today's world capitalist economy later, here it is necessary to stress the following from the standpoint of the matter we are examining.

Despite the fact that the cyclic synchronousness of the movement of capitalist production and of the world market already lias a history of 150 years, a line of refusing to recognise the general patterns of this movement still clearly dominates capitalist economics. When, however, the course of the world cycle is studied, its development is mainly considered as a kind of arithmetical sum total of the results of cyclic fluctuations within the national economies taken separately.

As a result it comes about that the specilic features of national development are, as it were, the sole decisive factor in the shaping of each world cycle. In that case the role of the system of international socio-economic connections and of the internationalisation of social production, which in essence convert these countries into links in a single economic chain, is patently underestimated.

Since the beginning of the 70s, it is true, because of the steady exacerbation of modern capitalism's international economic problems, an increasingly negative attitude towards sucli a 'narrowly nationalist' approach to analysis of the main world crisis processes (including its cyclic development) has been traceable in capitalist economic literature. Sometimes, while justly noting the close connection between these processes, and in particular between national overproduction crises, within the context of the capitalist economy, most capitalist economists as usual, however, prefer to identify their prime causes with the processes taking place at any definite moment in the sphere of the international market and political relations. They thus willy-nilly leave aside objective study of the system of international exploiter relations of production that underlies the deep internal contradictions of the world capitalist economy, which are inexorably undermining its main foundations.

The political dependence of most of the countries of the world on the main centres of capitalism has played an extremely important z'ole in the long history of the building up of this system. The structure of its unequal, exploiter relations was also moulded by that dependence in the imperialist stage, the crisis of which developed with particular force in the second half of this century as a consequence of most colonial countries' winning of national independence.

POSTWAR CAPITALIST WORLD ECONOMY

The growth of human resources in the much contracted postwar capitalist world economy has been unparalleled in scale. The population of all its countries increased by more than 1,350 million in 1950–80, which was more than double the corresponding figure for the first half of the century (620 million). As a result the total population of the capitalist and developing countries was nearly 3,000 million at the beginning of the 80s (as against only a little more than 1,600 million at the beginning of the 50s and 1,000 million at the beginning of the century). In other words, human 119 resources increased by 80 per cent between 1950 and 1980 within the non-socialist world, and by roughly 190 per cent since the beginning of the century.

These considerable changes were determined to a crucial extent by the 'demographic explosion', whose causes are linked with the break-up of the colonial system and advances in health protection. These points have been widely discussed in Soviet socio-economic literature, so we shall touch here only on those consequences of population growth that characterise its distribution in the principal regions of the capitalist world and are essential for understanding the shifts in the distribution of manpower among them, and consequently of the long-term trends in the development of their basic productive force.

Such an immense disproportion has not, by any means, been typical of the whole period of the development of capitalism's world system. Estimates based on UN demographic statistics indicate that even at the beginning of this century the growth rate of the metropolitan countries was patently higher than in the colonies and dependent countries of Asia, Africa, and Latin America, where 120 epidemics raged and hunger and high infant mortality prevailed. The expectation of life in the latter was about half that in the metropolitan countries. Since the war there has been a marked exacerbation of the demographic situation in the capitalist per centorld. The rates of increase began to rise in the former colonial world. In 1938–50 they averaged 1.4 per cent a year, in the 50s 2.3 per cent, and in the 60s and 70s 2.6 per cent. In the developed capitalist countries the rates were also rather higher in the first postwar decades, but again began to fall and in the 70s were around 0.85 per cent a year, *i.e.* about the average for 1900–1938.

POPULOUS CENTRE OF WORLD CAPITALISM

The most populous centre of world capitalism has always been and still is Western Europe. At the middle of the century rather more than half of the total population of capitalist countries lived there, but the growth rate in most West European countries has been very low in recent decades compared with other geographical regions, while emigration has been very high, which has deepened the trend already long apparent towards a reduction of the weight of Western Europe's population in both the capitalist countries as a group and the capitalist world as a whole.

The population of the North American area increased much faster than Western Europe's after the war. In the period 1950–81 its annual average growth rate was 1.30 per cent, which led to a certain convergence of the absolute size of the populations of the two main centres of capitalism. While the ratio between them was 1.7 : 1 in favour of Western Europe in the early 50s, it was 1.4 : 1 at the end of the 70s.

p One must remember, however, that the population of North America was increased in the first half of the century, and especially as a result of World War II, by a flow of immigrants from outside, including Western Europe, in addition to a rate of natural growth higher than in other developed capitalist countries. At the same time there was a high population growth rate in other capitalist countries, above all in the biggest of them, Japan, although it was also much lower than in the developing countries. A consequence of all these factors was a considerable decline in the weight of the industrial centres in the total population of today's capitalist world. The population of Africa (not counting South Africa) rose by 210 million, or more than doubled, in the same period, and its average annual growth rate began to exceed that of Asia in the 70s. These rates were not, however, limiting ones. Latin America set records of sorts, its population increasing by 120 per cent by the end of the 70s compared with the early 50s, with a mean annual growth rate of 2.75 per cent.

These major shifts in the ratio of population among the various groups of countries convincingly emphasise the need to make careful allowance for the demographic factors when studying the long-term development trends of the productive forces in the postwar capitalist economy. And they are of special significance for a sectoral analysis of the results of the main regions' economic development. At the same time these factors are extremely important for forecasting growth of the productive forces in any one economic system and on a global scale. The Soviet sociologist Arab-Ogly has justifiably commented that the tempestuous growth of population is one of the most important historical phenomena of our time as regards its remote consequences. It is facing both individual countries and mankind as a whole with a number of exceptionally complicated though in principle resolvable economic, social, and political problems. That is why there is an urgent need for substantiated and as exact as possible forecasts down to the year 2000. The compilation of such forecasts has now acquired universally acknowledged theoretical and practical significance. Demographic processes as a rule have immense inertia, which also makes it possible, strictly speaking, to make more or less substantiated forecasts of future populations.

In spite of the marked rise in tempos in various countries and years, the dynamics of population growth is limited in the principal regions by quite definite parameters, which provides the necessary premises for a more or less long-term estimate of the future distribution of manpower among them. A real effort has been made in UN demographic publications to tie up population growth trends somehow with certain outlooks for the subsequent course of development of production. One of its surveys remarked, for instance: The future size, structure and distribution of population are essential for any plan that involves food, housing, employment' education, health or other public services. This is also an era of increasing awareness: of man's unprecedented growth, of the interaction between this growth and the environment, and of the possible implications for the future. Investigations into these complex relationships have added to the demand for population projections.

For all the substantial differences in estimates of the future development of the international demographic situation, the majority of them quite validly, in our view, assume that one can hardly expect any extraordinary changes in growth rates in the really foreseeable term either in the capitalist or the developing countries compared with the postwar period, though some demographers' forecasts assume a certain lowering of these rates in coming years. In any case, however, there are adequate grounds for staling that the number of inhabitants of all the countries of the non-socialist world will increase by at least 50 per cent in the last two decades of the century (16 to 20 per cent in the first group and 56 to 60 per cent in the second). In absolute terms this means an increase in the population of the industrialised countries of roughly 120 to 150 million, and in the developing countries of 1,200 to 1,300 million.

These estimates, like all similar generalising forecasts, are approximate, but in themselves they are very important not simply for understanding the outlooks for the development of population but even more for clarifying their interconnections with the possible trends of the future growth and distribution of the productive forces in the countries at present forming the world capilalisl system. The postwar period has been marked by a further increase in the unevenness of the distribution of the main productive force in the developing world. Most of its inhabitants, and so of its manpower, has long been concentrated, as we know, in Asia, whose weight in the total population of the capitalist world has been steadily rising in postwar decades. Analysis of current trends gives grounds for sugg-esting that the Asiatic population will be around half in the mid-80s, and more than half in Ihe 90?, of the total population of the capitalist world within its present boundaries. But the majority of the liberated countries of Asia (in which more than 2,000 million people will probably be living at the end of the century, *i.e.* three times as many as in the middle of the century) are now the most backward areas, economically speaking, of the developing world. The population of Africa and Latin America is increasing even faster. These expected shifts arc undoubtedly fraught with serious socio-economic consequences for the fate of capitalism in those areas.

In coming decades jobs will have to be created in them for hundreds and hundreds of millions of people, but their agriculture is already unable to absorb any considerable part of the flood of new labour power. The objectively operating laws of agricultural production, moreover, are leading to surplus manpower being pushed out into the towns which, given capitalist relations of production, inevitably means a growth of poverty and proletarianisation of broad masses of the peasantry. That is one of the important distinguishing features of capitalism compared with the exploiter formations that preceded it. The law of faster growth of the urban population under the capitalist mode of production, Lenin stressed, 'does not and cannot operate otherwise than through the disintegration of the peasantry into a bourgeoisie and a proletariat'.

When that conclusion was drawn there was still no socialist economic system. Experience of its development has shown that for it advance of society's productive forces is inseparably linked with priority growth of the urban population, but it also indicates that the new mode of production, unlike the capitalist, makes it possible to avoid the extremely grave consequences of urbanisation for the broad masses objectively inevitable for all capitalist countries.

Since the war urbanisation has intensified to an unusual degree under the impact of the shifts in the structure of social production considered above. It has begun to have an ever more active effect on the character of the distribution of manpower in the world capitalist economy. In the 50s, 60s, and 70s the urban population of non-socialist countries more than doubled, so that the ratio of urban and rural population altered, with socio-economic and political consequences of no little importance. At the end of the 70s more than 40 per cent of the capitalist world's total population was already living in towns, while less than a third had done so in the early 50s. In the industrial countries this proportion rose to 75 per cent, and in the developing countries to a third, with fewer and fewer of the rural inhabitants available for work being directly involved in agrarian production. At the same time an increasingly large amount of manpower is becoming concentrated in the towns, a trend that has become stable not only in industrially developed capitalist countries but also in the overwhelming majority of developing countries.

There are, of course, considerable regional differences in this respect in both groups, largely associated with features of their preceding historical development. In the industrial centres the highest proportion of urban population is in North America, particularly in the United States, where more than 90 per cent of the total population was living in towns at the end of the 70s. There is an even more significant spread in the weight of the urban population in developing countries, from 15 or 20 per cent in Asia to 60 or 65 per cent in Latin America. One of the most characteristic trends in recent urbanisation is the rapid increase in the role of former colonial regions in the total urban population of the capitalist world in both absolute terms and relatively. At the beginning of the period reviewed the bulk of the urban population lived in industrial countries, while at the end already more than half of the townsmen lived in the agrarian and primary producing countries. The size of this population grew roughly twice as fast in the latter as in the former, and analysis of the long-term trend allows us to suggest that more than half of the population of the contemporary capitalist will be townsmen by the end of the century. At least twothirds of them will be living in the former colonies and semi-colonies of Asia, Africa, and Latin America.

A not unimportant feature of the long-term shifts in the distribution of population is its growing concentration in very big cities. The number of cities with more than a million inhabitants increased in the 50s and 60s alone by more than 50 per cent, while the number of their inhabitants more than doubled. Big cities have been growing much faster, moreover, in developing countries. Most of the biggest cities of the contemporary capitalist, world are now already located in developing countries. The chaos of capitalist urbanisation is leading to the formation of broad urbanised zones with multimillion populations in both groups of countries, which are drawing an ever bigger part of the population of adjoining towns and rural areas into their orbit, thereby intensifying the unevenness inherent in capitalism and anarchy in the distribution of society's main productive force.

Substantial shifts are also taking place in the geographical distribution of the rural population of the non-socialist world, which has grown in recent decades exclusively in the agrarian and primary commodity producing countries. In the industrially developed countries there has been an absolute fall in its numbers, while it has increased by more than two-thirds in the developing countries as a whole, and exceeded 1,600 million at the end of the 70s. In the principal economic centres the complete ousting of the petty producer from agriculture was completed, or almost so, in the postwar decades. Engels pointed out the objective inevitability of this in his time, remarking that the development of the capitalist form of production has cut the lifestrings of small production in agriculture; small production is irretrievably going to rack and ruin.

AMOUNT OF WORLD CAPITALISM'S RURAL POPULATION

In the majority of emancipated countries that have taken the road of capitalist agrarian relations, this process is developing with increasing force, but attempts to force its pace are still complicated by the low level of development of the other spheres of their economies, above all of industry. The proportion of the industrial countries in the total rural population consequently fell by half in the 50s through 70s, while it rose from less than 75 per cent in the developing countries to 90 per cent. As a result there was a considerable extension of agrarian overpopulation in the latter. This enormous overpopulation, which was long a heavy burden on their economies, still remains one of the most disastrous social 'side-effects' accompanying their economic growth under capitalism. An increasing amount of world capitalism's rural population is thus concentrated in developing countries, above all in Asia and Africa, where the rapidly growing towns are unable to absorb the villages' immense surplus of manpower. At the beginning of the 80s around two-thirds of this rural population lived in Asia and more than 20 per cent in Africa. The only major region where its proportion has begun to fall is Latin America, and there too, in fact, it still maintains its tendency to grow in absolute terms (increasing by more than 40 per cent in the period under review). In the coming decades the marked imbalance will apparently not be smoothed out, but will become rather more acute. In fact, all in any way valid long-term demographic forecasts envisage a progressive reduction of the rural population in capitalist countries and an inevitable growth of it in developing ones by at least a third by the end of the century.

The absence of sufficiently reliable estimates of the agrarian overpopulation in developing countries is a factor making it extremely difficult to determine exactly the scale and dynamics of employment in the whole world capitalist economy. The inadequacy, in turn, of the data on full or partial unemployment in industry and other spheres of production in most developing countries, and in some capitalist ones, is also an obstacle; so, too, are the constant upswings and recessions in employment through the effect of marked cyclic and non-cyclic fluctuations of the capitalist economy. Still, however, there arc composite indicators in the international statistics that enahle us to estimate, with the necessary degree of comparability, the growth trends in the long term of that part oi' the population that could be involved in the reproduction process. This type of indicator primarily includes estimates of the scale of growth of the gainfully employed population, analysis of which helps us arrive at the characteristics of the development trends of productivity for the different groups of countries.

The imbalance in the postwar growth of the able-bodied population between the two groups of the capitalist economy has increased immensely. Over the twenty years of the 60s and 70s, for example, their numbers increased twice as slowly in the developed capitalist countries as in the developing ones. Consequently there was a tendency for the proportion of the former in the total gainfully employed population of the capitalist world to decline. From the fact that the decisive mass of the population that will begin to work at that time in both groups of countries already exists, we can assume that around three–quarters of the gainfully employed population of today's nonsocialist world will be concentrated in the now developing countries of Asia, Africa, and Latin America in the mid-80s, and around i'our-iifths by the end of the century. But this labour force will still obviously be relatively unskilled in the main while the most skilled part will be concentrated as before in the economically developed centres.

In that connection the problem of the migration of various categories of workers between countries with different levels of development will be aggravated. In many industrial countries there has been a fast growing shortage since the war of national personnel in the mass trades that do not call for any degree of professional skill. The shortage began to be covered more and more by importing labour from economically less developed countries.

In the 70s this shortage constituted around ten or twelve million, and may (according to some estimates) increase several times over again by the end of the century. As for the developing countries the inherent tasks of their socio-economic development will steadily lead, in the first place, to an increase in their need for well-trained, skilled personnel, a need that cannot be wholly met simply from local resources. The possibilities for attracting specialists from outside, however, are greatly limited by the increasing need for such personnel in the developed countries themselves, and also by the comparatively high standard of their pay there. It is not fortuitous that a mounting portion of the West's funds for 'technical aid' are spent on the specialists sent to developing countries, and on training students.

At the same time, monopoly capital, utilising the laws of capitalist competition, is everywhere broadening the 'import' from developing countries of the most talented and capable part of their intelligentsia and skilled specialists. The 'brain drain' from Asia, Africa, and Latin America has become a marked feature of the international migration of labour. The spontaneous flow of surplus labour between separate developing countries has also become unparalleled in scale in recent decades.

THE EVALUATION OF CAPITAL PROJECTS

Institutional economics is a policy-oriented discipline in which the significance of economic inquiry is a function of its relevance for problem-solving, including those dilemmas faced by governments. Virtually every central or regional government has some overview on how it sees its distinct economic sectors, such as that of water, developing over time. Within sectors, programmes will be elaborated and programmes require projects, the cutting edge of economic change.

The provision of government finance for a country's river-basin, water-supply and sanitation activities is always likely to face a budget constraint. It follows that, in choosing one set of initiatives demanding government resources, others inevitably have to be rejected or at least postponed. For example, Kinnersley has referred to the competition between irrigation projects and the provision of basic water supplies in some countries. The question then arises, from the point of view of the country as a whole, of 'What is the optimum combination of initiatives to select in any given year?' An entire branch of economics, known as social cost-benefit analysis (SCB A), can assist in answering this type of question. SCB A concerns itself with the evaluation of capital projects. There is no assumption that the technique is a substitute for politics; rather, it interacts with politics.

The literature on social cost-benefit analysis is enormous. It then continues with issues specific to SCBA that the economist must confront: the choice of the discount rate; the use of shadow prices; the exclusion of transfer entries; and various estimation problems. Thereafter, the ground shifts to consider whether project analysis can have any relevance to demand management.

PROJECT DEFINITION

A project's characteristics are that it usually has a well-defined geographical location, it is an activity with a specific starting point at which time investment costs are incurred and during the period of the project's life a series of outputs are produced. The lining of a set of irrigation canals is an example. Gittinger notes: 'Usually [the project] is a unique activity noticeably different from preceding, similar investments and it is likely to be different from succeeding ones, not a routine segment of an ongoing programme.' The project cycle can be classified in alternative ways. Here is an example that highlights the process as an economic activity:

- *Identification.* In this stage the project is conceived and its broad outlines agreed. This might come about as a result of a sector survey of broader scope.
- *Preparation and analysis.* A more detailed homing-in on project definition occurs with feasibility studies of alternative approaches, covering economic, financial, technical and organizational aspects.

- Ex ante *evaluation.* The project is evaluated in economic terms prior to its launch, on the basis of forecast expenditure and income on capital and current account. *Ex ante* social cost-benefit analysis is appropriate at this stage.
- *Implementation.* This breaks down into the gestation period, when the bulk of the investment costs are incurred and subsequently the working period of the project until it comes to an end.
- Ex post *evaluation.* This takes place at the end of the gestation period and at any point during the project's productive life or at its end. In the last case, all the data are retrospective, in contrast with the forecast and prospective intelligence of *ex ante* evaluation. The value of *ex post* appraisal is in developing subsequent strategies, programmes and projects on the basis of the wisdom of hindsight; it also can expose systematic errors in *ex ante* evaluation.

The general objectives of evaluation in economic terms are to test whether the returns on projects exceed their costs, to rank projects with respect to their economic efficiency and to help policy-makers choose a set of projects, within the financial constraints facing them, which contribute most effectively to the country's economic development in a manner consistent with its social and environmental values. In the context of this book, SCBA is a technique primarily intended to raise the efficiency of government expenditure in support of water infrastructure programmes, particularly in respect of irrigation and drainage in agriculture.

In its preparation and analysis phase, the project cycle should include an appraisal of the impact of an investment proposal on specific actors, such as the inhabitants of an area to be flooded for dam construction purposes, the civil engineering industry, regional government, farmers and manufacturing firms. This is vital to understand who is likely to benefit or to lose from the project and what their anticipated behavioural response may be. But the method for the evaluation of capital projects concerns itself primarily with the costs and benefits of the activity from the point of view of the nation as a whole. For this reason, project costs and returns are considered in terms of the real economy, *i.e.* the flow of real resources used up in order to produce flows of real outputs or services. This holistic rather than partial approach is why the term *social* cost-benefit analysis is used.

THE NET PRESENT VALUE APPROACH

A starting point for understanding the technique of *ex ante* SCB A is to see it as a means to judge whether any single project will (or will not) make a positive contribution to a country's gross domestic product (GDP). This requires that-over the full life of the project, *i.e.* the gestation period plus the working period-one identifies the annual real flow of goods and services produced *with* the project in place, in comparison with the real flow of output *without* the project. This will be called the incremental real flow. At the same time we identify the incremental annual real flow of resources used in order to produce that output.

These identified physical quantities of outputs and inputs, in the form of goods and services, can then be given a common standard of measurement, in terms of their market prices. A spreadsheet may be drawn up of the annual value of the incremental real output in row:

- The value of the incremental real costs in row
- These two rows are hereafter referred to as the project's benefit and cost flows. The difference between benefit and cost in each year is the net benefit flow and is recorded in row
- A water project deemed to have a full life of 25 years, starting at the beginning of year 0 and terminating at the end of year 24.

Where the project has clear external effects in GDP terms that the producer neither pays for nor receives payment for and where these are quantifiable in price terms, such positive or negative externalities should also be included in the benefit, cost and net benefit flows. Kinnersley has pointed to the frequency of negative externalities from water projects:

In economic terms, the most common hazard is the way in which rivers spread damage caused at a particular point. Activities that pollute the river or divert part of its flow usually affect water quality or flow for some distance down stream. The water may thus be made unsuitable for other uses down stream, but often without would-be users down stream being able to trace who or what caused the pollution and thereby claim compensation. Thus, the polluting upstream activity, assuming it is a business, may be said to be escaping costs of waste disposal and preventive measures against pollution which it should normally have to provide for and reflect in its selling prices.

In the early period of a water infrastructure project, the net benefit flow is likely to be negative, when capital account outlays fail to be matched by the stream of benefits. Thereafter, net benefits may remain positive through to the end of the project's life, although major rehabilitation outlays in the mid-years could modify this general rule. In all cases we would expect the identification and valuation of benefits to be even more difficult than those of costs.

An important issue arises immediately. Should the market prices we are using for the valuation of the incremental real flows be constant or out-turn prices? Constant prices are those ruling in one specific year and are applied to the outcomes of both that year and all other years considered in the analysis of the project. Out-turn prices are the ruling level of prices in each separate year and, of course, vary over time. The standard practice in the field is to use constant (usually base year) prices. The argument is that a general price inflation does not, in comparison with zero inflation, indicate any difference in the real value to society of outputs or inputs. This rule notwithstanding, the planning of a project's financing should consider the likely level of general price inflation, so that an adequate budget is obtained to cover the costs of the gestation period.

However, if there is a strong expectation of a *relative* shift in the prices of some outputs or inputs, quite aside from any general movement in the price index, these would be reflected in the calculation of the net benefit stream. Here the assumption is that relative change implies a shift in the value that society, through the market, places on those commodities.

A second issue that deserves to be addressed at an early stage concerns the mode of project finance. Should SCBA as a general practice embody in its calculations the sources of the money that will fund the project during its gestation period? Gittinger argues persuasively that this should not be the case.

He suggests that one assumes all financing for a project comes from domestic sources and that all its returns go to domestic residents. This is consistent with the general objective of SCBA defined in terms of projects' contributions to GDP. Project analysts almost universally accept this convention of splitting the evaluation of individual projects from their specific financing sources. It is perfectly consistent with identification of possible financing sources in the preparation and analysis phase of the project cycle that precedes or runs alongside *ex ante* appraisal. The convention also neatly side-steps the fungibility enigma.

By this stage, we should have completed a spreadsheet. Let us suppose that *N* is positive, *i.e.* the total value of the stream of incremental real output *(B)* exceeds the total value of the flow of incremental real costs (C). In such a case, is it correct to assume that the project should be approved? Can SCBA really be that easy?

Unfortunately it is not, for two distinct reasons. First, such a procedure would ignore the timing of the benefit and cost flows. Second, such a procedure would provide no basis for choosing between projects, either where two (or more) are mutually exclusive for technical reasons or where a budget constraint imposes a limit on the total number of projects that can be financed- the starting point. The term *project interdependence* will be used to refer to either of these two situations.

With respect to timing, in project evaluation it is universally accepted that each annual benefit and each annual cost entry should have its value reduced, *i.e.* it should be discounted, the later it occurs. The mathematics of discounting is set out below. The accounting conventions are that the total project period is divided into years, that rates of growth or contraction are expressed in annual compound rates and that all receipts and payments in each year occur at the beginning of that year.

Suppose US$1 is received at the beginning of year 0 and that in each successive year the sum received increases at a constant compound rate of 5 per cent. Then, after 5 years this series of payments in dollars and cents is as set out.

Thus, the entry for each receipt in any given year *t* is equal to:

$$1.00(1.05)'$$

More generally, for any given proportionate rate of increase i, expressed as a decimal, we can write the receipt as:

$$1.00(1 + i)'$$

This is a compound growth series. But when we discount the future, we are suggesting not growth into the future but, so to speak, contraction back from the future. We argue that an entry *t* years after the present (the start of year 0) needs to be reduced proportionately each year to calculate its present value.

One divides, instead of multiplying, by the exponential term. As a double check, note that in row 1, the receipt from year 4 when divided by $(1.05)^4$ gives a present value of 1.00.

The decision to discount each and every entry after year 0 should be carried out, as equation 6.3 above indicates, through dividing it by $(1+d)^t$ where, again, *d* is the selected discount rate (such as 5 per cent) and *t* is the year in which the entry appears.

These individual discounted values of the benefit, cost and net benefit streams are added up and recorded. They will be written here as B^*, C^* and N^*. N^* is known as the net present value (NPV) of the project. Where the NPV is positive, the project can be approved on economic grounds, provided project interdependence does not exist.

The apparently laborious task of dividing dozens of cell entries by $(1+d)^t$ can be simplified with a calculator, with a published discounting table, or with an NPV function available through standard spreadsheet computer software. In the approach above, all costs and benefits in the first year are undiscounted, since they are assumed to occur at the beginning of year 0. [Note that $(1+d)^0=1$.] Alternatively the first year of the project is counted as year 1 (not year 0), all entries are deemed to have been incurred at the end of each year and, once again, we have the convenient expression $(1+d)^t$ to use to discount, where *t* is the project year in which the entry falls. Here, first-year receipts and outlays *are* discounted.

Next it is necessary to consider the difficulty that, in common experience, not all project proposals are appropriately considered in their own right-as independent of each other. In the first place, on technical grounds two (or more) proposals may be mutually exclusive, most obviously when each is an alternative means by which a water infrastructure investment can secure a given policy objective. In the case of mutually exclusive projects, the solution is simple: calculate the NPV of each option and choose the proposal with the highest NPV.

In the second place, projects may not be appropriately considered as independent because a single programme budget is available for their financing and the budget is insufficient to fund them all. The question is then posed: 'On economic grounds what is the optimum subset of proposals to which finance should be provided?' In this case, for every proposal (including all mutually exclusive projects) we calculate the net benefit-investment ratio (NBIR). This is simply done by dividing the full project life into two: the gestation period, when the net benefit flow is consistently negative and the working period, when the net benefit is, in most years, positive.

The NPV for each period is separately calculated and that of the working period divided by that of the gestation period. The negative sign of the denominator is ignored; absolute values are to be used here. All projects are then ranked from top to bottom, using the net benefit-investment ratio. Lower ranking mutually exclusive projects are eliminated and then all the remaining projects are approved off the top until the budget constraint blocks any further approvals.

What we have done in the second case-that of financially interdependent choices-is to consider the key constraint to be the budget funds available during the gestation period and then to rank all projects in terms of the productivity (in NPV terms) of these funds. This maximizes the total NPV generated by the use of the funds available. The net benefit-investment ratio can also be considered as a means not of excluding projects but of ranking them in terms of their starting date.

When applied to irrigation and drainage projects, the net benefit-investment ratio is a valuable output-input ratio and I shall refer to it as *E*10, yet another measure of irrigation water efficiency.

Discounting the Future

Up to this point, the NPV approach to the evaluation of capital projects has been presented in terms of its most basic elements. But if SCB A in this form is to be an acceptable technique for project decision-making, some extremely important areas of debate must be introduced. These are classified under four headings: discounting, shadow prices, transfer entries and estimation errors. It was argued that the timing of the benefits and costs of a project should be handled in the evaluation process by reducing these values in proportion to the time delay before the benefit or the cost is registered. The method for doing this was described as the division of each value by $(1+d)^t$, where d is the selected discount rate and t is the year in which the entry appears. But no reason was given as to why the timing of benefits and costs is significant, nor why they should be handled by means of a discounting process.

The justification for this approach falls not within the field of economics but of psychology. There seems to be an almost universal human propensity to regard a real benefit offered now as superior to the same real benefit offered later in time. Human life is beset with risks of injury, loss and death. That has always been the human condition and is doubtless the origin of a rather rational desire to receive now, when the probability of receipt is greatest, rather than later, when such probability is perceived to be less. This mind-set is an example not of human frailty but of sweet wisdom distilled from the bitter lessons of experience.

Project evaluation, rightly in my view, recognizes this psychological propensity by the use of the discounting technique in respect of project outputs. This still leaves us with the difficult question: 'At what level should the time rate of discount be set?' Since the evaluation of capital projects is here placed within a national framework, setting out from the concept of the GDP, one is clearly seeking a value of *d* that is appropriate to society as a whole and so one refers to this rate as the social time rate of discount (STRD).

To grasp the reasoning here, it may be of some help to draw a parallel with the physicist's concept of the half-life of radioactive material. In applying the arithmetic of discounting to handle the psychology of risk avoidance, we are suggesting that a real output (or cost) available now with a value of US$1.00 has proportionately lower value today if it is forecast to be available only in a year's time and a still lower proportionate value for the following year. And so on. Thus today's value of that projected real output contracts or decays at a constant rate the later the time at which the benefit is forecast to be available, rather like the radioactive strength of some elements decays at a constant proportionate rate over time. For any given value of a real output (or cost) at the beginning of year 0 and for any specific measure of the social time rate of discount, it is easy to say how many years will elapse before that original value is deemed to have decayed by one-half. This number of years I shall call the half-life of the value of incremental real output implicit in any measure of the STRD.

The definition of the half-life associated with a specific social time rate of discount needs careful specification. For any real benefit valued today at US$1 if it is available today, but valued today at only US$0.50 if it is forecast to be available in *h* years time, then *h* is the half-life. The half-life of the value of output, cost or net benefit is given for every integer value of the STRD between 1 per cent and 15 per cent. The results are dramatic. At a value of 1 per cent, the half-life is as long as 70 years. But at 15 per cent, the half-life is only 5 years. The absolute value of the half-life falls substantially up to an STRD of about 5 or 6 per cent, where it averages close to 13 years.

This figure may help decision-makers to set the appropriate rate for project evaluation in their own countries. In my view the whole thrust of discussions on sustainability in recent years suggests that high values of the STRD are inappropriate, as they produce short half-lives-they markedly devalue the medium- and long-term effects of project investment decisions. This point of view has been strongly urged by Lipton (1992) in the context of the high real rates of interest that were current in the 1980s and 1990s. A benchmark STRD of 5 per cent per year may be widely appropriate, which would give a half-life for net benefit entries of about 14 years. An alternative to the NPV approach is used by organizations such as the World Bank and most other international financing agencies, as Gittinger reports. This calculates the internal rate of return (IRR) of a project, *i.e.* the discount rate at which the discounted value of incremental real output precisely equals the discounted value of incremental real costs. A cut-off rate is then set and all independent projects above this value are deemed worthy of approval. Note that when this IRR is based on the inputs and outputs in financial terms it is called the financial rate of return (FRR); when based on inputs and outputs in economic terms we have the economic rate of return (ERR). There are problems with the IRR approach. First, the cut-off rate is said to represent the opportunity cost of capital in a country, but in fact no one knows what this is. Moreover, the concept of the opportunity cost of capital rests on the assumption that its use would lead to the approval of all possible investments scoring above the cut-off rate. Yet such an outcome is entirely inappropriate, since choice on that basis ignores the huge importance of environmental and social effects in correct decision-making. Second, the ERR cut-off rate is used as a crude ranker of projects, but it is recognized that the ERR should not be used to rank projects at all. Third, when constant prices are being used, chosen cut-off rates are typically in the range 8-15 per cent. If the appropriate STRD is much lower, as I have suggested, then the use of such a high cut-off rate for the ERR introduces a bias against sustainability, favouring projects with lower (but earlier) undiscounted returns.

The advantages of using a net present value approach based on the social time rate of discount are that:

- It forces decision-makers to make choices about the appropriate social time discount rate in their society;
- Once this is set, no unjustifiable bias in favour of the short term takes place;
- It gives a practical ranking for interdependent projects, using the net benefit-investment ratio, which recognizes in a pragmatic way the opportunity cost of capital funds.

The use of the IRR at a high discount rate can have powerful effects on the routines of design. My friend Stephen Allison writes:

After I finished my Ph.D. I worked for Harza Engineers in Chicago and, in the course of the first six months, wrote an internal guide entitled 'Designing for a 15 per cent IRR'. That was the Bank's (almost only) criterion in the mid-sixties. I made the point that if one did a preliminary feasibility study which yielded 13 per cent, the thing to do was seek places where less equipment or construction materials could be used that were less durable and of lower capital cost. Early replacement, or higher O&M costs, would not 'hurt' the project as much as higher capital outlay! Before moving on, a few brief words are in order on SCBA's hard-worked cousin-social cost-effectiveness analysis (SCEA). Broadly speaking, whenever two or more projects achieve the same outcomes but in different ways, they are mutually exclusive on technical grounds. Project analysis limited to such a package of alternatives can be carried through merely by calculating which option has the lowest discounted social costs, *i.e.* which is the most cost-effective from the point of view of society as a whole. Winpenny for example, cites the case of irrigation canal lining in southern California in the use of SCEA.

TYPICAL OF CAPITALIST ECONOMICS

It is typical of capitalist economics, in contrast to Marxist, to try and put markets and market relations at the centre of its analysis and not production and the class relations immanently inherent in it. The former are looked upon as determinant or self-sufficing, allegedly developing of themselves and exerting a dominant influence on the production process. The idea of the primariness of the sphere of circulation, with very substantial differences in the evaluation of the main moments of the functioning of economic systems of any scale, has dominated capitalist economics throughout its history. True, its original theoretical postulates are now being criticised more and more often by many leading Western economists, people like Samuelson, Galbraith, Myrdal, Erhard, Leontief, Balassa, and Tinbergen.

Still, this idea has continued in fact to permeate capitalist literature in one form or another since World War 11, whether it deals witJi capitalism's internal or world economic problems. In their detailed work on the growth of the international economy, Kenwood and Lougheed declare the central sphere of their study to be that of circulation, since, as they write:

The exchange ol goods and services is the means through which independent economic units enter into economic relations with one another and become part of local or national economic community. An exchange passes beyond a country's boundaries, national economic systems become parts of broader regional, continental or world economy.

Prof. P. T. Ellsworth stresses, in his textbook, which was widely used in the United States and Great Britain in the 50s and 60s, that the very term 'international economies' indicates 'that it concerns itself with the economic relations between nations'. Such a posing of the matter is quite typical of present-day Western economic literature.

When investigating capitalism's world economic problems, capitalist economists do not, of course, ignore theoretical analysis of the international aspects of the production process, and study many of them very thoroughly, particularly the dynamics of its productive forces, comparations of the level of economic development of various countries and regions, problems of the international division of labour, capital exports, production and technological competition, integrational processes within the various international economic groupings, migration of labour, and so on.

All these processes, however, arc willy-nilly analysed from ideological positions that pursue definite class interests of the capitalists. The world economy is studied mainly from the standpoint of the determining effect of the trends of world supply and demand on production, and of opportunities for effective employment of the market mechanism of state monopoly methods of regulating international economic relations. Objectively the main job of this approach is 18 to conceal the antagonistic essence of the international relations of production under capitalism. Capitalist economists' endeavour to survey the world capitalist economy out of the context of its organic link with the patterns of development of the corresponding exploiter relations of production, and to represent it simply as a mechanically formed aggregate of national markets within the context of the international exchange of goods and services, also serves the same purpose. On this basis capitalist political economy has time and again exerted great efforts to revive the gimcrack idea whose essence is that modern capitalism is still ahle to generate adequate conditions for future 'harmonious' development of the world economy and in the long run to eradicate its deepest inherent defects and contradictions.

Marxist-Leninist economics considers this matter from fundamentally different theoretical positions. It substantiates in detail the proposition that the world capitalist economy, like a commodity economy of any scale (whether local, regional, or national), includes such inseparably interconnected and jointly subordinated elements as production, markets, and a system of socio-economic relations arising from a corresponding division of labour between economic units and producers. Although market factors of exchange and distribution figure primarily on the surface in capitalism's international economic interconnections, it is not they, however, but the social relations in the production process, that play the basic role when the antagonistic, class nature and decisive trends of development of capitalism's world economic system are investigated.

The capitalist mode of production (Marx wrote) is... a historical means of developing the material forces of production and creating an appropriate world market and is, at the same time, a continual conflict uetwcen this its historical task and its own corresponding relations of social production.

This conclusion has acquired special methodological significance for analysing the deep-seated crisis processes in the modern world capitalist market. It provides an opportunity for a scientific systematising and comprehensive investigation of the objectively operating patterns and irreconcilable contradictions of capitalism's international market relations, which stem from the socio-economic nature of this mode of production.

INDUSTRY GROWTH IN DEVELOPING COUNTRIES

As for industry, its growth in developing countries has still not led to any real qualitative shifts in the liquidation of age-old industrial backwardness. The struggle to lay an industrial foundation for their national economies is still more or less, as a rule, in the initial stage. In spite of a narrowing of the gap in per capita industrial output, they now produce dozens of times less product than the average for capitalist countries.

The most powerful, technically advanced branches of production, moreover, that can more efficiently exploit the fruits of the present scientific and technical revolution are concentrated in the latter. That is affecting the dynamics of productivity in the industry of both groups of countries. In the 60s and 70s its annual growth rates in the economic centres of capitalism were double that in the developing countries.

There has been a quite stable tendency towards a gradual, tliough slow, rise in the proportion of circulation and services in the GDP of the developing world, which rose perceptibly in 1900–80. As a result of the rapid expansion of this sphere its weight in the GDP of the Asian and African regions rose by a quarter and nearly a third respectively in the 50s through the 70s. It rose a little as well in Latin America on the whole and constituted more than half of that region's GDP at the end of the 70s. In other words, there was a certain convergence of structural proportions in the sectors of non-materialproduction of the main developing regions, but the historical underdevelopment of these sectors still remains a distinguishing sign of the economic backwardness of the vast majority of the former colonies and dependent countries of these regions.

The proportion of developing countries in the aggregate production of the services and circulation sphere of the world capitalist economy is beginning lo grow. At the start of llio 80s it was 10 per cent against 11 per cent in the early 50s. The per capita gap between them and the developed countries, it is true, not only did not in fact diminish but even widened a bit on the whole, and was estimated at 15.5:1 in the late 70s, *i.e.* was considerably wider than in the sectors of material production.

For the main sectors of the circulation and services sphere this gap has not built up in the same way; in practice it has always been narrower in trade. But it is in that sector, as in agriculture, that the capitalist countries continued steadily to outpace the developing countries in per capita indices of growth, so that, although the proportion of the latter in the total value of the home trade of the capitalist world lias risen since the war (from one-seventh to one-sixth), there has been another trend in per capita terms While the value of per capita trade (in constant 1975 prices) in the GDP of capitalist countries was 12 times as much as in developing countries at the beginning of the 60s, it was 16 times as much at the start of the 70s.

This trend above all determined the effect in the circulation and services sphere of the tendency towards a further widening of the general economic gap in per capita GDP between the two groups of countries. The intensification of this trend also to no small extent was promoted by processes taking place in the sectors producing so-called services.

Developing countries' backwardness in this respect has long been deeper than in the other sectors of the sphere of social production we are considering. At the turn of the 50s and 60s it corresponded approximately to the level of their industrial backwardness. At that time less than one-tenth of the value of the services created in all capitalist countries was produced in developing countries, and the gap per capita was nearly 20:1.

In recent years there have been changes in various directions in these ratios. The rapidly increasing dynamics of the growth of services in developing countries has led to their share rising in absolute terms to one-seventh at the end of the 70s. But in per capita terms the gap continued to widen for a long time, and in the early 70s could be expressed as 19.5:1, which was mainly due to the backwardness in this 175 area of the economically less developed and more densely populated countries of the African and Asian regions.

In the second half of the 70s alone, especially during the world cyclic crisis of 1974–75, a long-term consequence of which was a marked slowing of the growth rales of services in the centres of capitalism, did this gap begin to narrow a bit; at the beginning of the 80s it was expressed by a ratio of 17.5:1 (in 1975 prices). In other words, it has remained wider than in industry. The extreme backwardness of the services sphere governs several of the paramount aims of the long-term socio-economic strategy being developed by developing countries today in their striving to attain the maximum possible speeding up of the rates of social progress in present-day conditions and to raise the standard of living of their populations.

The only major group of services and circulation sectors in which developing countries managed to get a relatively stable increase in the dynamics of per capita production over that of the developed countries in the postwar decades was transport and communications; the grand total of per capita production in them increased by a factor of 3.4 against 2.4 in the developed countries.

There was a corresponding ri se (from 11 to 16 per cent) in developing countries' weight in the aggregate value of the product of transport and communications in the world capitalist economy. In per capita terms their mean annual rates of development have been higher of late than those of the industrial regions, so that in the area of the infrastructure, and in industry, there has begun to be a relatively new and apparently promising trend towards a certain narrowing of the gap in per capita indices, though it still remains extremely wide. If the present dynamics of the growth of production and population is maintained the gap should be halved at least on an average in the coming two or three decades in each of the main regions of the developing world, but its ultimate closing will undoubtedly still remain an urgent task for a long time.

At the same time there was a clearly increasing unevenness in the development of transport and communications in the developing countries, which, as the figures given above show, 176 was organically characteristic of all the other sectors of tho circulation and.services sphere. Its dynamics will also undoubtedly be a close function of the degree and scale of development of material product ion and of the population grovyth rates in the various regions and countries of the developing world.

On the whole, however, one can say with confidence that as the productive forces rise thc weight of this sphere in the social production of these countries will gradually increase although it will still lag greatly behind the corresponding average indicator for the industrialised countries for several decades.

The wide gap between the developing countries and the centres of capitalism in all the decisive areas of economic activity consequently remains one of the objective realities of our time. Formed long ago, during the building of the world capitalist economy on a colonial basis, it now characterises many of the most important features of that economy, and of tho whole system of its postwar international relations.

The developing countries of Asia, Africa, and Latin America, given the existing differences in their levels of sociopolitical and economic development, arc inevitably coming up against a need to deal with several common problems, vitally important for all developing countries, on both an international and a national scale. They include the very intricate set of each country's national problems, but ultimately tho struggle to cope witli all of them, taken together is objectively directed to eliminating the pernicious surviv' als of the colonial period from the affairs of human society" and to emancipating the developing countries from neoco, lonialism and continuing capitalist exploitation.

At the same time, as will be clear from the facts adduced above, the gap in levels of economic development remains extremely wide, in spite of considerable progress by the developing countries in coping with these problems, and moreover retains a tendency to widen in several determinant indicators, above all in gross product per capita. The relations of inequality and rapacious exploitation of the natural and human resources of the developing countries by expatriate monopoly capital prevailing in the world capitalist economy are furthering maintenance of this trend, even since the break-up of imperialism's colonial system.

The capitalist social system itself not only bears responsibility for the socio-economic backwardness of the peoples of the former 177 colonial world, but the exploiter system of international division of labour created by it largely continues today to fetter their productive forces. The basic patterns of the world capitalist economy still operate in the direction detrimental to the developing countries.

The striving of the developing countries to spread the liquidating of colonialism to the economic sphere, to put an end to their exploitation by the industrial powers of the West, and to achieve creation of the necessary external and internal conditions for attaining a level of development in the foreseeable future corresponding to the needs of today, is quite justified. At the same time all the nations of the earth without exception have an interest in the speediest attainment of these aims. The struggle for a steady acceleration of socio-economic and cultural progress in the former colonies and semi-colonies is therefore becoming an inseparable part of the fight of all progressive mankind for detente and the consolidating of lasting peace, for further development of the scientific and technical revolution for peaceful purposes, and for a favourable solution of the other global problems of today on which the whole subsequent course of world development to an enormous extent depends.

In his day Karl Marx concluded that mankind thus inevitably sets itself only such tasks as it is able to solve, since closer examination will always show that the problem itself arises only when the material conditions for its solution are already present or at least in the course of formation.

In the present historical situation more favourable objective conditions are being created than ever before for the countries of Asia, Africa, and Latin America that have liberated themselves from the imperialist yoke, or are in the course of doing so, for a successful struggle to overcome the backwardness inherited by them from the past in all sectors of 178 the economy and oilier areas of social affairs.

But every kind of obstacle is still being put in llio way of this bard, long, stubborn light with the world capitalist economy by the former colonialists and expatriate monopoly capital. In these conditions the developing countries are uniting their efforts more and more firmly in the anti-imperialist movement to establish a new international economic order.

The Soviet Union and other socialist countries are supporting these efforts in every way possible and demonstrating their solidarity with the progressive aims of the struggle of the peoples of the developing world for a better future. As the Central Committee's Report to the 26'th Congress of the CPSU stressed: the CPSU will consistently continue the policy of promoting cooperation between the USSR and the ncwly-freo countries, and consolidating the alliance of world socialism and the national liberation movement.

The long-term trends reviewed in this chapter thus not only help us better to understand some of the general results of the changes that have already taken place in the structure of the distribution of the productive forces of the two groups of countries of the postwar capitalist world, but also to distinguish sqveral objective premises for analysing the probable outlook for subsequent structural shifts leading to aggravation of the internal contradictions and crisis phenomena in the capitalist economy. Study of these trends obviously calls for a need to allow in every way for the effect on them of the spontaneously operating laws of the cyclic development of the capitalist mode of production.

PRODUCTION DISTRIBUTED AMONG THE DEVELOPING COUNTRIES

The most important trends in the changing postwar structure of the periphery's social production have been analysed above. How, however, was this production distributed among the developing countries in the first postwar decades? As will be clear from the graphs the growth rates of aggregate production were very fast in Latin America, where most of the countries had achieved national independence and taken the road of independent capitalist development long before break-up of the colonial system began. As a result the postwar years have been marked by a gradual raising of the role of Latin America in the aggregate product of the developing world, from 40 to 44.5 per cent, so that it is now greater than that of the Asian region although the latter had nearly four times its population at the end of the period surveyed.

This growth did not come about through the contribution of all Latin American countries, but occurred on a background of a marked imbalance of their economic development and was mainly determined by a few leading countries, primarily Brazil and Mexico, for which much higher growth rates were typical. At the end of the 70s more than 60 per cent of the aggregate product of the region was produced in these two countries alone, whereas they produced around 40 per cent of it at the beginning of the 50s. At the same time certain countries (Haiti, Honduras, Barbados, etc.), still caught up in heavy chains of dependence on expatriate commodity monopolies, remained among the most economically 159 backward appendages of the world capitalist economy, with the usual low growth rates typical of semi-colonial countries. A steady deepening of the unevenness of Hie intor-regional. and especially of the intraregional, distribution of production has become typical of the various groups of commodityproducing countries in the other continents. In Africa, where colonialism succeeded in retaining key positions in many areas for a number of years in the postwar period, the course of economic development has proved to be relatively slow, so that the weight of the continent in the aggregate social product of developing countries has had a clear tendency to fall, despite the fact that the total GDP of African countries has risen substantially.

As national independence has been consolidated and African countries have thrown off the yoke of colonial exploitation, however, their growth rates, it must be noted, as in other areas of the developing world, have gradually risen. They have been highest, as a rule, in the Arab countries in North Africa, where the positive consequences of the breakup of the colonial system made themselves felt much earlier than in Tropical Africa. The rapid increase in income from exploiting their oil wealth, moreover, provided some of them with favourable material conditions for their struggle for quicker economic development.

There has been an ever wider difference in levels and rates of development in Asian countries, where around two-thirds of the developing'world's population lives. Most of Ihem, which were in forced dependence on a handful of European powers during nearly the whole history of the forming of the world capitalist economy, were able to win national independence at the start of the postwar period. In the 50s they had already begun an active fight for accelerated growth of their previously stagnating productive forces. The mean annual growth rates of the GDP of all the developing countries of Asia reached 4.5 per cent in that decade, 5.1 per cent in the 60s, and around 6 per cent in the 70s. This accelerated growth was due to no small degree in recent times, however, to the 'oil boom' in the Near and Middle East. The real volume of the GDP (in value terms) increased more than sixfold in that region in the 50s through the 70s, and was more than one-third of the GDP of all the developing countries of Asia, although its population at the end of the period was only 7 per cent of the continent's total. At the same time the GDP of all the other developing countries in South and South-East Asia, which have immense human resources, rose less than threefold. If we examine these processes through the prism of the data on population growth, the generalisations made above, stemming from a comparative description of the regional dynamics of production, obviously need to be made more precise, and corrected. The successive changes in the indicators of per capita GDP for the separate regions and countries enable the scale of the shifts in the geographical distribution of their productive forces to be brought out more clearly.

In almost all the groups of countries in the table, there is a quite marked tendency, albeit extremely varied in dynamics and significance, towards a differentiation of population growth rates, in addition to a sharpening of the differences in dynamics of production. This trend characterised one of the inevitable consequences of the break-up of the colonial system, and could not help having an essential influence on the ratio of economic potentials, and consequently on the mutually interlocked processes of economic development. In the Asian region a particularly wide gap has been formed in the dynamics of per capita production of the aggregate social product between the oil-producing countries of the Near and Middle East and the overwhelming majority of the countries of South and South-East Asia.

Similar processes have been occurring in Africa, for all its special features. The growing uncvenness of development has been deepened there, even more than in Asia, by considerable differences in rates of population growth. In some cases these differences, it is true, have encouraged a convergence of the dynamics of per capita production in the separate groups of countries. That is confirmed, in particular, by comparison of the lines of postwar development of North and Tropical Africa. Population has grown faster in the former, which has led to a certain levelling out of the aggregate indicators of per capita growth of production used in the table. At the same time there have been quite different trends within each of these regions.

In the biggest North African country, Egypt, for example, there was an obvious slowing of the growth rates both of production and of population in the 60s and 70s. The annual rates of per capita growth of GDP in those years were 1.5 per cent in Egypt and only 0.2 per cent in Sudan. In other words, they were lower than in many of the countries of Tropical Africa whose economies on the whole still suffer maximally from the burden of harmful survivals of the colonial period compared with other developing countries. Although some of the new states south of the Saharajiave achieved a relatively high growth of production (Kenya, Gabon, Ivory Coast, Nigeria, Malawi, etc.), most of them still have extremely slow and very unstable rates of development. In quite a few countries, moreover, the main indices of the dynamics of GDP growth have been lower in recent decades than the growth of population (Burundi, Chad, the Central African Empire, Uganda, Upper Volta).

TRENDS IN THE DISTRIBUTION OF PRODUCTION

The long-term trends in the distribution of production in the former colonial regions are primarily governed by the resultant tendencies of the development of a few of the biggest countries (as regards area, population, and natural 164 resources). The ratio of their economic capabilities is therefore important in a generalised description of intraregional processes, including the contradictory consequences of the 'demographic explosion' in the Third World. In Latin America, for example, GO per cent of the inhabitants and nearly two-thirds of the aggregate GDP of the whole region were concentrated in Brazil, Mexico, and Argentina. Among the 'Big Three' production developed at the lowest rates in Argentina in the postwar years, but the comparatively slow dynamics of its population growth led to its per capita production, for example, exceeding Brazil's in the 60s and Mexico's in the early 70s. It is typical that the economies of several small countries (Uruguay, El Salvador, Paraguay, Haiti, Honduras, and others) wore below the average of the developing countries of South and SouthEast Asia as regards the same indicators in those years, not to mention of bulk of the emancipated countries of Africa. The examples given, whose number could be greatly extended, very clearly confirm that there has been a steady sharpening of the unevenness of economic development of the former colonies and semi-colonies in recent decades.

This irresistibly intensifying process, however, in no way refutes the need (in addition to the study of the features of the economic growth of each country) for a comprehensive systems approach to analysis of the summary trends of their interconnected development within the whole developing world. Such an approach helps, on the one hand, to bring out the objectively operating trends in which the extremely contradictory and isolated processes and phenomena characterising the shifts in the distribution of the productive forces of the separate countries and their regional groups are ultimately reduced, and on the other hand, gives the possibility of better understanding and appreciating the specific features of their development.

In this connection comparative analysis of the long-term changes in the industrial structure of the social production of countries in Asia, Africa, and Latin America is becoming of paramount importance. As these countries are industrialised this process will undoubtedly hec07iie more and more pronounced.

Its effect gives grounds for supposing that even before the end of this century the share of agriculture in the GDP of most of the developing countries of Asia and Africa may fall by 50 to 50 per cent and reach the level attained in Latin America al the beginning of the 80s. At the same time the proportion of agriculture in the GDP of the Latin American region as a whole will also probably diminish, gradually approaching the present average level of the industrial capitalist countries (around 4 per cent). An inevitable consequence of that will be the pushing of increasing numbers of the rural population into the cities.

Another very important trend determining botli the current structure of changes in the production of the developing countries and their probable future structure has become the general though not uniform rise-in the role of industry in their GPP. The scale of its growth, moreover, has been quite regularly higher in the economically more backward countries. Thus the proportion of industrial output in the aggregate product for the period considered in the graphs increased in the African region by 120 per cent (from 12 to 26 per cent), in the Asian region by almost 100 per cent (from 14 to 27 per cent), and in the Latin American region by almost a third (from 24 to 31 per cent). As a result of the increasing imbalance of this growth, a trend towards a certain convergence of the significance and place of industry in the general structure of these regions' production has been noted in the postwar period. The proportion of industry in the GDP of countries in South and South-Easl Asia and in Africa was 50 per cent less than in Latin America at the beginning of the 50s; at the beginning of the 80s the gap had noticeably closed and was estimated now at 1:1.3 for the former and 1:1.2 for tho latter.

This trend will intensify, in all probability, in the future, and characterise one of the tendencies of developing countries' uneven economic growth. It can be expected that the weight of industry in the GDP of most Asian and African] countries will come close to tho present level in the leading Latin American countries in the next two or three decades, while the latter will reach the average level of the present-day industrial centres of capitalism.

These very general estimates provide the dnta needed, but only the initial data, for studying the significance and place of industry in the regional economies. They naturally smooth over tho vast differences in structure and sectoral dynamics of the industrial production of the separate regions and countries.

A more and more concrete analysis of the separate industries is a central task of the part that follows, but here we must bear in mind that these processes of industrial development in agrarian and raw material producing countries are largely linked with rapid growth of mining, whose output is mainly sold on the world capitalist market, and in fact is still really controlled to one degree or another by expatriate monopoly capital. The rapid growth of the mining of minerals, and in particular of fuel, has played no small role in the regions being considered in the postwar rise in the proportion of industry in the gross product.

It would not be legitimate, however, to conclude from this that the steady raising of the role of industry in the gross product of developing countries is more or less wholly 168 linked with the working of mincrals.

The rapid development of their manufacturing industries in recent decades has had an ever increasing effect, especially the development of heavy industry, the volume of whose production rose on the whole by almost 150 per cent in 1968–80.

At the beginning of the 80s the heavy industries of all developing countries were producing 30 per cent more output in value terms (in added value in 1975 prices) than the light industries. There is little doubt that their basic industries will also grow al faster rates in the next few decades, which will further a gradual raising of their role in industrial production.

For a comparative estimate of the long-term trends bringing out both the general and the particular in the production sphere of thè main developing regions of capitalism, it is essential to compare the mean annual growth rates of their main industries (including per capita rates) over some considerable period. Because of the inadequacies of the economic statistics of developing countries it is very difficult to bring out these growth rates in summary form (in contrast to developed countries).

The comparison can only be attempted on the basis of very approximate calculations, which allow us to characterise the most general trends of the sectoral shifts with a certain degree of accuracy. At the same time they can be used as a starling point of sorts for studying the specific features of the development of production in the separate countries and groupings in each of the regions.

In agriculture, which is still in a slate of deep, protracted depression in the developing world as a whole, Latin America and most of the countries of the Near and Middle East have displayed relatively faster growth, so that a line towards an increase in their share in developing countries' total production of farm produce has begun to show., with a corresponding drop in the role of South and South-East Asia and Africa.

As a result they have become further differentiated as regards per capita farm production. By the beginning of the 80s the Latin American countries on the whole considerably surpassed the countries of Asia and Africa in this respect. It is unlikely that there will be any closing of this gap in the near future.

The fight to eliminate the extreme backwardness of agriculture will undoubtedly remain one of the most pressing socio-economic problems of most of the countries of the former colonial world in the coming decades. It is the slow development of their agriculture, compared with population growth, that is the decisive cause of the consistent tendency manifested since the war towards a widening of the gap in GDP per capita between the developing and developed countries of the capitalist world.

In the 60s and 70s, however, this trend began to be more and more actively countered by processes developing in most of the other sectors of the economies of the former colonial periphery of capitalism, and primarily in industry. The international statistics clearly indicate that the per capita growth rates of industrial production in economically most backward regions of capitalism's periphery have noticeably exceeded the industrial centres in per capita growth rales of industrial production. In the developing countries of Asia per capita industrial production increased by a factor of 4.3 from 1950 to 1978, in Africa by 3.5, and in the developed capitalist countries taken together by 2.6. The countries of Latin America also showed a rather faster growth in this respect, on an average, than the main economic regions of capitalism.

In present conditions there has begun lo be a tendency towards a certain closing of the gigantic, gap in per capita industrial production formed under colonialism, together with 171 a gradual increase in the weight of the agrarian and raw material producing countries in the industrial production of the non-socialist world. In Hie very early 50s this gap was estimated at 1:35 for Asian countries, 1:32 for Africa, and 1:5.5 for Latin America. By the end of the 70s it had been clearly reduced and was expressed by the following approximate ratios: Asia 1:24, Africa 1:27, and Latin America 1:5.3. In the middle of 70s the substantial slowing of industrial growth rates in the developed countries connected with the greatest postwar cyclic crisis of the capitalist world economy fostered a consolidation of this trend.

The general course of this process leads to th,e conclusion that it is acquiring a quite stable, irreversible character in Ihc present situation. Being a natural consequence of the break-up of the colonial system, it has already begun to have (and in all probability will continue to have) a growing impact on the steady deepening of the crisis of the unequal system of international capitalist division of labour built up on a colonial basis.

At the same time the facts adduced confirm the extreme complexity and difficulty of the problem of overcoming the age-old industrial backwardness of the former colonies and semi-colonies, especially with their constant exploitation by expalriate monopoly capital. The solution of such a major and objectively inevitable hislorical task will obviously require several decades and will continue for a long time to be one of the urgent problems ol the light of the peoples of the liberated countries for final liquidation of the heavy legacy of the colonial period and imperialist domination of their economies.

To reduce their backwardness by even two-thirds from the present mean per capita industrial production of the developed countries, for instance, it would take the countries of South and South-East Asia around a quarter of a century at the rates of real industrial growth of the 60s and 70s, Africa more than a quarter of a century, and Latin America around 30 years. Then, however, per capita output of industrial goods in each of these regions would the respectively one-eighth, a ninth, and about a third below the level already reached by Ihe advanced capitalist countries at the end of the 70s. If that level remains unchanged in the future, then in that hypothetical case the way out for most developing countries would have lo be a 15-fold or 16-fold expansion of the volume of industrial production within the present-day world capitalist economy.

These figures cannot, of course, serve as a generalised criterion of a quantitative, let alone a qualitative, estimate of the future parameters and goals of the industrial development of all the developing countries of Asia, Africa, and Latin America. In fact, however, they help determine only the scale of the industrial backwardness of developing countries. It inevitably follows from them that the struggle for a cardinal solution of the problem in the present transitional epoch will give rise to an urgent need to develop a long-term strategy and-strictly scientific approach to the global outlook for industrialisation of the former colonies and semi–colonies on quite other principles of social and world economic relations compared with those historically built up by capitalism.

The postwar processes in the agrarian and industrial spheres have consequently had a different effect on the resultant ratio of per capita GDP in the groups of countries under consideration. In the agrarian sphere they have fostered a widening of the economic gap, and in the industrial sphere a narrowing. But since the role of the agrarian sector in the GDP of developing countries is more significant than in the centres of capitalism, and the industrial sector less significant, the gap between them in per capita output of the main sectors (agriculture, industry, and building) has also continued to widen in recent decades, and for Asian and African countries was more than 10:1 at the beginning of the 80s against 8:1 in the first postwar years. In other words, at the turn of the decade to the 80s, per capita production in value terms (in comparable prices) was much lower in developing countries compared with capitalist ones, than at the turn of the 40s and 50s.

8

Commercial Agriculture

Although elements of the above ideal type of a competitive food system could be found in parts of the United States at various times, it never did exist in many areas of the country. For an individual farmer, the question was not how many firms were involved in the different stages of the food chain across the country, but rather how much of this commodity chain was accessible to his (sometimes her) farm. AS farmers moved west, one of their major problems was how to transport their products to markets in the eastern cities. The government, wishing to promote increased industrialization, also perceived the problem and subsidized the construction of transportation systems, especially railroads. This often made the farmer dependent on a monopoly which could exploit him/her because of the unequal balance of economic power. If a farmer had access to only one railroad, the power relationship certainly favored the railroad. That farmer faced a monopoly regardless of how many other railroads existed in the country. Thus, many farmers faced the issue of monopoly control of capital from the time they became commercial farmers and began to be dependent on a single transportation system to move their products to the market.

Railroads in some parts of the country needed the business of farmers, but they had access to hundreds, if not thousands, of farmers. They were not dependent on any single farmer. Their only concern was that farmers might be successful in organizing a united stand against the railroads. Frank Norris' agrarian populist novel, The Octopus: A History of California (1901), was about a handful of farmers who attempted such a united stand, rising up in direct rebellion, only to be dispatched by the railroads. The whole history of the farmers' movement is largely about the unequal power balance between farmers and the railroads, and, more generally, between the farmers and all the "middlemen" they depended on for transportation, markets, and a host of inputs such as credit and farm equipment.

In southwestern Minnesota, for example, the selling of water fowl to cities to the East and later the selling of agriculture products, especially grain, all depended on the railroad for transportation. In the case of grain, the farmers also depended on the elevators (large silos) to store and to transfer it from farmers' wagons to the train. Many of the transnational corporations (TNCs) of today, like Cargill, exercised economic power in many of the local markets in which they began operating. Most farmers' movements were not successful in establishing alternative economic systems or firms which benefited the farmers, but with the help of legislation to encourage farmer cooperatives, there were some successes.

DIVERSIFICATION AND CROP INCOME STABILITY

The effectiveness of diversification as a self-insurance strategy. Stability is measured in a relative context and is identified with a low coefficient of variation for household crop income. Crop income is measured in two ways, gross returns and net returns to family owned resources which include land, capital, family labour and management. All non-marketed outputs and inputs are valued at their respective opportunity costs in measuring gross and net returns.

Of the 180 cultivator households in the VLS sample, 134 have data on crop income for each of the five years from 1975-76 to 1979-80. The majority of the other 46 households leased out their land for one or more cropping years; a minority left the sample and were replaced.

Therefore, there are 134 cultivators for whom coefficients of variation of household gross and net crop returns are estimated from 1975-76 to 1979-80. CVs are calculated for each household across the five cropping years. It was also expected that crop income stability is a function of some of the resource base determinants. Moreover, differences in crop income stability among farmers can also be attributed to fluctuations in the resource base across cropping years. To assess the impact of diversification on crop income stability, it is important to control for shifts in cultivated area, irrigated area, and land quality over time. This is particularly true for regions such as Sholapur where tenancy is prevalent (Jodha, 1981).

Descriptive information on the CVs' of the dependent and independent variables. The size of the CVs' at about 40 to 50 per cent is another indication of the riskiness of dryland agriculture in the semi-arid tropics of peninsular India. Fluctuations in gross returns are somewhat less than in net returns. Somewhat surprisingly, both average gross and net returns are less stable in the more heavily irrigated Mahbubnagar region compared to the dryland Sholapur and Akola regions.

It is believed that crop income stability can be traced not only to the level of the resource base but also to its intertemporal variation. Because crop and plot diversification are themselves functions of the same resource base viewed in the context of association rather than causation.

It is perhaps heroic to try to explain inter-farm variation in crop income stability within a narrow ecological region based on only five year data. Nonetheless, some of the hypothesized determinants behave as expected and have statistically significant estimated coefficients. Crop diversification is negatively and strongly associated with crop income instability in the more dryland Sholapur and Akola regions. A 10 percent proportional increase in the crop diversification index is associated with 4.8 and 13.0 decrease in the CV of net crop returns in Sholapur and Akola respectively. At the margin, crop diversification is about three times more effective in stabilizing net returns in rainfall assured Akola than in drought-prone Sholapur.

In Mahbubnagar, plot diversification contributes positively to crop income stability. In contrast, in the Akola and Sholapur villages, the results mildly suggest that plot diversification is associated with greater crop income instability. Production conditions for a crop in these more dryland villages must be quite narrowly defined as diversification across space with the same commodity does not dampen fluctuations in income. This is particularly true in Sholapur where field diversification is positively correlated with gross and net return instability. These results support Walker and Singh who found land sharecropped in Sholapur was of inferior quality and was more likely to be the site of crop failure than owner-operated land. Plot diversification appears to be a resource management strategy rather than an action that offers scope for crop income protection.

As expected, the CVs' of the resource base determinants are for the most part-the exception being land value in Mahbubnagar-positively associated with crop income instability. Larger farms having more gross cropped area are characterized by greater crop income stability in the three regions. However, the impact of farm size on crop income stability is not large.

Evaluated at their arithmetic means, a 10 percent increase in gross cropped area leads on are average to about a 2 percent reduction to the CV of net returns. The evidence with respect to irrigation is not as clear, but it weakly suggests that greater access to irrigation is accompanied by greater instability in net returns particularly in the Akola villages where irrigated land accounts for only 3 percent of the gross cropped area. Besides following inter-regional differences were also observed:

1. The impact of irrigation on crop diversification and income stability hinges upon cropping conditions and the existing level of irrigation. When an expansion in irrigation takes place for rainy season paddy production in a region where many farmers have some access to irrigation, irrigation leads to more specialization in paddy production. In contrast, when an investment in well irrigation occurs in a multicrop predominantly dryland region, it enhances opportunities for further crop diversification particularly in the postrainy season. More profitable production opportunities in turn translate into greater relative variance in net crop returns income and increased crop income instability. The net effect of irrigation on crop diversification and income stability depends on the source of irrigation, ecological location, and level of growth of irrigation facilities.

2. Crop and plot diversification fluctuate more in more drought prone regions like Sholapur. Crop diversification appears to be more sensitive to agro-climatic determinants than plot diversification.
3. Different regional ecological conditions offer differential opportunities for risk averse farmers to diversify. Therefore, it is not surprising to find that risk attitudinal differences vary regionally in their effect on diversification. Where they are significant, differences in risk attitudes do not exert a sizeable influence on decisions on diversification; their proportional impact is about the same as differences in draft power availability.
4. Crop diversification effectively imparts stability to net crop returns in the dryland farming Sholapur and Akola villages. In the more highly irrigated Mahbubnagar villages, income instability and crop diversification are not significantly associated.

The implications of these conclusions are threefold. First, at present levels of technology it appears that increasing land fragmentation associated with plot diversification does not represent a significant cost in terms of income foregone or a material benefit in the form of enhanced crop income stability. Thus a prospective land consolidation programme would not compromise the risk-bearing capacity of farmers, nor would it significantly increase average crop income of farm households at existing levels of technology. These issues require closer scrutiny within a more rigorous household economics framework.

Secondly, the emergence of crop diversification as a strong force in protecting net crop returns in the more rainfall-assured Akola region decreases the demand for government risk-reducing policies, such as crop insurance, because farmers can rely on self-insurance through diversification to enhance crop income stability. These results further underscore the emerging consensus that the potential benefits of crop insurance programmes will be greater in more risk-prone, dryland regions like Sholapur for three reasons. First, yield variability is the predominant source of risk in these regions (Barah and Binswanger, 1982), yields are more likely to be positively covariate across farmers which augments the efficiency of a homogeneous area approach to dampen fluctuations in crop income and third, there is less scope for crop diversification to be an effective self-insurance measure.

Lastly, an investment in irrigation may have a negative effect on relative crop income stability when instability is measured with respect to the coefficient of variation. The farm-level effects of irrigation may be markedly different from what obtains at the district, state or national levels.

ECONOMIC GROWTH AND AGRICULTURE

Economic growth is an increase in per capita income. Agricultural growth is necessary for economic growth. Nearly every high-income country has a highly productive agricultural sector, and agricultural growth is critical in the process of economic growth. Among other things, economic growth involves the production of more goods and services, and a much wider range of goods and services, than before. Economic growth agriculture's share of the economy declines with economic growth. As per capita income increases, percentage of labour force in agricultural production declines. Agriculture's share of Gross Domestic Product (GDP) also declines. GDP is a widely used measure of national income. Across states at a given point in time, or at a single country over time, the relationship between per capita income and agriculture's percentage of the labour force and GDP looks. Agriculture tends to decline more quickly as a percentage of GDP than as a percentage of the labour force.

POLICIES TO MARKET ACCESS

Poor infrastructure, limited resource base, lack of information, lack of or inadequate support institutions and poor policies affect the livelihoods of small-holder farmers. Other barriers include market standards, limited information, requirements for large initial capital investments, limited product differentiation, and handicapping policies. While almost any of the farm produce sells at the village level market, consumers discriminate against produce that is comparatively inferior, hence farmers have, over time, adapted to selling only that which will sell.

POLICIES RELATED TO GLOBALIZATION

The effect of globalization on any country depends on that country's level of economic development, structures in place during the implementation stage, flexibility of its economy. Globalization has three dimensions: multiplication and intensification of economic, political, social and cultural linkages among people, organizations and states at the world level; the tendency towards the universal application of economic, institutional, legal, political and cultural practices; and the emergence of significant spill-over from the behaviour of individuals and societies to the rest of the world. Due to the interrelation of the various dimensions, policies made in one country are bound to have effects on another. With globalization, comes liberalization of markets. The food security threat caused by liberalization is due to dumping of heavily subsidized produce in developing states and pre-mature exposure of upcoming industries to genuine competition from producers in developing and developed states. In addition, most profits are repatriated by transnational companies reducing the potential for poverty reduction to direct employment alone. In most cases, the pay is low because the national policies do not protect the small-holder.

HANDICAPPING POLICIES

Poor policies have greatly affected the food security in India. This is mainly due to uneven development within states based on political grounds.

Certain regions are developed preferentially for political reasons at the expense of others. The problem arises when the focus on policies, structures and institutions is put above that of the people themselves. When policies are not inclusive in their design they tend to handicap the exempted lot by providing barriers.

One such way in which this may take place is uneven development within states where certain regions are preferentially developed for political reasons at the expense of others. Polices that promote monopolistic competition for the large-scale industries hurt the cottage and small industry. When we fail to provide safety nets for vulnerable groups, we doom them to destruction.

CHANGE PROCESS AFFECTING AGRICULTURAL PRODUCTION AND LIVELIHOODS

The lack of transparent, timely and reliable crop and livestock marketing information is seen by many as one of the greatest challenges to the development of the agriculture industry in East India. Without access to information about agricultural products prices at different regional markets, farmers are unable to identify which points-of-sale offer the best prices for their agri-products.

For example, in Mbirikani, Kajiado County, Kenya, Maasai pastoralists once relied on middlemen to gain market information. However, a marketing information system has been developed by the Livestock Information Network and Knowledge System (LINKS) to support livestock producers and traders. Now the farmers use mobile phones to ask for livestock prices in Emali, Mombasa and Nairobi, and this knowledge significantly improves their bargaining power with livestock traders.

INCREASING MARKETING OPPORTUNITIES

Given the high dependency of pastoralists' livelihoods on the sale of livestock and livestock products, LINKS designed and implemented an Information Communication Technology (ICT) infrastructure to collate data on livestock sales and prices from a network of different markets for dissemination using SMS messages. This has been a success and has been used to develop a National Livestock Marketing Information System (NLMIS) in Kenya. NLMIS relies on a network of County livestock marketing officers and data monitors to transmit data between markets and the database.

The field officers are also trained to download, analyse, and summarize the information to transmit to pastoralists and traders, including information about prices and volumes of cattle, camels, sheep and goats at livestock markets. The market information is also made available online, and may be downloaded for printing and sharing with livestock communities with no internet access. The information is also shared by e-mail, posted on billboards at market places and can be requested through SMS.

Vibrant Markets

This change in communication has also transformed undeveloped markets in other value chains in agriculture. The availability of marketing information through SMS and information boards, competition among agri-products traders has increased, and this has improved and stabilized prices paid to farmers. Women have also benefited by providing animal health services, fodder and small loans to the traders. Empowering producers to negotiate for better prices has equipped market agencies and communities with appropriate tools and information to plan for and respond to changing incomes, and consequently has improved livelihoods.

Policy

The success of the marketing system has equipped planners and policymakers with the tools to track price trends, which are helping them, improve decision-making and devise appropriate interventions to mitigate the effects of deteriorating terms.

IT FOR AGRICULTURAL PRODUCTION AND MARKETING

IT is playing an important and vital role in agricultural production and marketing. IT allows farmers to save time on order and delivery and getting feedback. In the existing competition, there is a need to rapidly attract new customers as well as retain existing customers.

In order to take the real status of agricultural production and marketing, there is an urgent need to develop the following items:

- *Farmers' Crop Database must be Managed*: The database includes the kinds of crops, the size of cultivated area, time of harvest and yield. Farmers or the extension personnel transmit those data via the Internet to database server. Further, information provides the farmer with an important instrument for decision making and taking action.
- *Crops Information Service System should be Created:* This system analyses the crop data to create some statistical tables. Farmers can access these statistical data by browsing the homepage and make their production plan. Changes within the structure of agriculture will probably have an impact on the selection and types of acquisition of software and other integrated systems made by the farmers.
- *Production Techniques and Information Enquiry System should be Created*: This system integrates the production techniques and information, which are developed by experimental agricultural institutes and agricultural improvement stations. Farmers can find out relevant production information through this enquiry service system.
- *Production Equipment's Enquiry Service System should be Created*: This system gathers information from the companies of seeds and crop

production equipment to build the production equipment's enquiry service system. At the same time, allow relevant companies to access this system and enter their own data. Therefore, farmers can order the needed items through this system.

Information is critical to the social and economic activities that comprise the development process. Development economy has witnessed for revolutions in agricultural, bio-technological, industrial and information technology. Good communication system and information system reinforce commitments to sustainable productivity. The Government of India is giving more thrust on agriculture, food and information technology sectors towards achievement of economic reforms to achieve high growth rate in production in the years to come.

The National Agriculture Policy announced addresses the challenges arising out of economic liberalization and globalization. It seeks to actualize the vast untapped growth potential of Indian agriculture, strengthen rural infrastructure to support faster agricultural development, promote value addition and secure a fair standard of living for farmers and farm workers. The National Agriculture Policy lays emphasis on the use of Information Technology for achieving a more rapid development of agriculture in India.

In pursuance thereof, the Department of Agriculture and Cooperation (DAC) has formulated information technology (IT) Vision 2020. This vision *inter alia* envisages that:

- Information relating to agriculture sector would be available to the ultimate users—the farmers—for optimizing their productivity and income;
- Extension and advisory services making use of information technology would be available to the farmers on round the clock basis;
- The tools for information technology will provide networking of agriculture sector not only in the country but also globally and the Union and State Government Departments will have reservoirs of data base; and
- The long term vision on "Information Technology in Agriculture Sector" is to bring farmers, researchers, scientists and administrators together by establishing "Agriculture on-line" through exchange of ideas/information.

In future, information technology will reduce the cost and time of information system. IT will bring new information services for agricultural development that will enable the farmers to have much greater control over the information channels.

AGRICULTURAL IMPLICATIONS OF IT

Information technology provides answers to a number of questions to the farmers. For example, what are the benefits of more irrigation? Is it cost-effective to apply additional chemicals? When is the best time to sell crops or buy inputs? With improved record-keeping, more detailed cost analysis and more sophisticated marketing strategies, farmers are making better decisions and earning higher profits.

The Internet is increasing communication and business opportunities within the agricultural community, which previously operated in the relative isolation of rural areas. Farmers, agricultural researchers, cooperatives, suppliers and buyers use the Internet to exchange ideas and information, as well as to conduct business with each other. Machinery, seed chemicals and other types of agricultural products can be purchased and sold online. People can search for jobs and employees. It is to be noted that, the farmer is in no position to use IT directly.

The literacy levels, language barrier as most of the application software are predominantly in English, cost of computers, poor communication infrastructure make it impossible for individual farmers, particularly small farmers to directly adopt IT. This calls for institutional effort to harness to create the necessary IT based services to farmers. But in India, one prominent problem is that most of the farmers own small holdings, this seems to be difficult. In this situation, it may be made possible by adopting the corporate farming system, which is the need of the hour with advent of new agricultural policy.

By taking up corporate farming, a group of farmers can install a computer and any educated young man from that group can undergo training of how to browse the Internet. He can provide the farmers current commodity, analysis reports on world markets and trade for different commodities. Food market overviews provide valuable information about some of the most important export market. IT can help to provide the information on the likely price distribution of key commodities over the coming years. Such information helps farmers and traders make decisions on when and in what ways to market their grain. Whether, to sell at harvest or store on-farm in anticipation of higher later in the season. When combined with enterprise budget data, the information can also be used in deciding which crops to produce in the coming season.

In order to encourage farmers to obtain best possible price, information on various agricultural output markets is also being provided. The objective of this activity is to provide status of price at different markets to facilitate farmer to move his produce to the market where he can expect better price. The entire exercise will not be useful unless necessary arrangements are made to ensure that the farmers utilize this facility. The contribution of information technology in bringing down costs, increasing efficiency and improving productivity and thereby contributing to the bottom line needs no special emphasis.

In the fertilizer marketing context, IT can play a major role in efficient sales operations, checking the marketing costs, safeguarding market share and providing efficient customer services. A well conceived IT set up can endow decision makers at all levels with better reflexes to effectively respond to market conditions. IT helps producers monitor and respond to weather variability on a day-to-day basis.

Solar-powered weather stations in the field can be hooked up to a farmer's computer to relay information about current air and soil temperature, precipitation, relative humidity, leaf wetness, soil moisture, day length, wind speed and solar radiation. Producers use the Internet to monitor prices quickly and as often as they like. Farmers from around the world can exchange ideas, post questions and get answers about specific topics. Thus, it is said that the importance of information technology in the field of agriculture is emerging.

THE COMMERCIAL GRAIN-FARMING AREA

This area covers the outer region in the south and the east of European Russia, the steppe gubernias of Novorossia and the Transvolga. Agriculture is distinguished here for its extensive character and the enormous scale of the production of grain for sale. If we take the eight gubernias of Kherson, Bessarabia, Taurida, Don, Ekaterinoslav, Saratov, Samara and Orenburg, we shall find that in 1883-1887 the net crop of cereals (not including oats) for a population of 13,877,000 amounted to 41.3 million chetverts, *i.e.*, more than one-fourth of the total net yield of the 50 gubernias of European Russia. The crop most commonly sown here is wheat—the principal export grain. Agriculture develops here fastest of all (by comparison with the other areas of Russia), and these gubernias relegate the central black-earth gubernias, formerly in the lead, to the background: Thus there is a *shifting* of the principal centre of grain production: in the 1860s and 1870s the central black-earth gubernias were ahead of all the rest, but in the 1880s they yielded priority to the steppe and Lower Volga gubernias: their production of grain began to *diminish.*

This interesting fact of the enormous growth of agricultural production in the area described is to be explained by the circumstance that in the post-Reform period the outer steppe regions have been *colonies* of the central, long-settled part of European Russia. The abundance of free land has attracted an enormous stream of settlers, who have quickly increased the area under crops. The extensive development of *commercial* crops was possible only because of the close economic ties of these colonies with central Russia, on the one hand, and the European grain importing countries, on the other. The development of industry in central Russia and the development of commercial farming in the outer regions are inseparably connected and create a market for each other. The industrial gubernias received grain from the South, selling there the products of their factories and supplying the colonies with labour, artisans, and means of production (timber, building materials, tools, etc.). Only because of this social division of labour could the settlers in the steppe localities engage exclusively in agriculture and sell huge quantities of grain in the home and particularly in the foreign market. Only because of their close connection with the home and foreign markets could the economic development of these localities proceed so rapidly; and it was precisely capitalist development, for along with the growth of commercial farming there was an equally rapid process of the diversion of the population into industry, the process of the growth of towns and of the formation of new centres of large-scale industry.

As to the question of whether the growth of commercial farming in this area is bound up with technical progress in agriculture and with the creation of capitalist relations, that has been dealt with above. We saw how large the areas cultivated by peasants in these localities are and how sharply capitalist relations manifest themselves there even within the village community. We saw that in this area there has been a particularly rapid development in the use of machinery, that the capitalist farms in the outer regions attract hundreds of thousands and millions of wage-workers, with huge farms created on a scale unprecedented in agriculture, on which there is extensive co-operation of wage-workers, etc. We have little left now to add in completion of this picture.

In the outer steppe regions the privately-owned estates are not only distinguished occasionally for their enormous size, but are also the scene of farming on a very big scale. Above we made reference to crop areas of 8, 10 and 15 thousand dessiatines in Samara Gubernia. In Taurida Gubernia, Falz-Fein owns 200,000 dess., Mordvinov 80,000 dess.; two individuals own 60,000 dess. each, "and many proprietors have from 10,000 to 25,000 dessiatines". An idea of the scale of farming can be obtained, for example, from the fact that in 1893 there were 1,100 machines (of which 1,000 belonged to the peasantry) haymaking for Falz-Fein. In Kherson Gubernia there were 3.3 million dessiatines under cultivation in 1893, of which 1.3 million dess. belonged to private owners; in five uyezds of the gubernia (without Odessa Uyezd) there were 1,237 medium-sized farms (250 to 1,000 dess. of land), 405 big farms(1,000 to 2,500 dess.) and 226 farms each of over 2,500 dess. According to data gathered in 1890 on 526 farms, they employed 35,514 workers, *i.e.*, an average of 67 workers per farm, of whom from 16 to 30 were annual labourers. In 1893, 100 more or less big farms in Elisavetgrad Uyezd employed 11,197 workers (an average of 112 per farm!), of whom 17.4 per cent were annual, 39.5 per cent seasonal, and 43.1 per cent day labourers. Here are data on the distribution of crop area among all the agricultural undertakings in the uyezd, both of private landowners and of peasants.

Thus, a little over 3 per cent of the peasants (and if we count only those who cultivated, 4 per cent) concentrate in their hands more than a third of the total area under crops, for the tilling and harvesting of which masses of seasonal and day labourers are required.

There is no need, apparently, to comment on these data. We have had occasion to observe that the area described is the most typical of agricultural capitalism in Russia—typical not in the agricultural sense, of course, but in the social-economic sense. These colonies, having developed with the greatest freedom, show us what relations could and should have developed in the rest of Russia, had not the numerous survivals of pre-Reform life retarded the development of capitalism. The forms, however, of agricultural capitalism, as will be seen from what follows, are extremely varied.

COMMERCIAL RELATIONS OF PRODUCTION

We are all familiar with the classical story of how capitalism came to dominate industrial production and how capitalist relations of production swallowed up the individual artisanal producer. We recognize the power that the capitalist mode has to infiltrate and finally transform other forms of the organization of production and exchange. We sometimes think that the power of that transformation is so great that all of the significant action already occurred in the past, at least in Europe and North America, and was essentially over by the end of the nineteenth century. In the society we inhabit, it is a fait accompli, whose dynamic we can only understand by reconstructing the past because it is not happening around us. On second thought we realize that the transition was still in progress until very recently in a few skilled domains like medical care and entertainment, where individual artisans were able to ply their trade throughout most of this century, but these fossils of early capitalist relations seem exceptional because of their requirement of special talents or of necessary skills acquired by long training. But the view that the transition to mature capitalism is essentially over except at the margins of the main body of commodity production is clearly wrong, because it ignores an immense sector of basic essential commodity production, agriculture, which is still in the throes of the transition.

PENETRATION OF CAPITAL IN AGRICULTURE

The penetration of capital into agriculture has been a long process of a different form than the classical case of industrial production usually exemplified by cloth weaving in the eighteenth and nineteenth centuries. Indeed, on the surface agriculture would seem to have been resistant to capital. After all, despite a 72 percent drop in the number of of individual farm enterprises in the United States from 6.7 million in 1930, there are still about 1.8 million independent farm producers today. This means that even though only 6 percent of these establishments account for 60 percent of the total value of farm production, there are over 100,000 separate enterprises producing more than half of all the value of the output. In the industrial manufacturing sector the four largest enterprises account for an average of 40 percent of value produced, and even in a highly differentiated product like clothing the top four companies produce over 15 percent of the value.

There has also been a major increase in the proportion of farmland that is leased to farmers who also own their own land. Roughly 55 percent of farmland is now operated by owner-renters who are for the most part small producers. Finally, despite the conventional wisdom that corporate farming is taking over, the proportion of farms and farmland operated by managers representing absentee owners has remained about 1 percent since the beginning of the century.

Thus if we are to look for evidence of the capitalist transformation of agriculture, we will not find it in the classical industrial model. We do not find a concentration of more and more productive capacity in the hands of a very small number of farmers, employing a large wage labour force carrying out its tasks under close supervision and according to a tightly controlled schedule. There are, of course, some examples of a factory-like labour process in farming, especially in the harvesting of fruits and fresh vegetables, and these are often pointed to as evidence of a capitalist transformation to factory farming. However, the vast majority of farm enterprises do not employ a large labour force, but more typically have one or two hired laborers, usually for only part of the year.

In analyzing the process of the capitalist transformation of agriculture we must distinguish between farming and the agrifood system. Farming is the physical process of turning inputs like seed, feed, water, fertilizer, and pesticides into primary products like wheat, potatoes, and cattle on a specific site, the farm, using soil, labour, and machinery. The failure of classical capitalist concentration in farming arises from both financial and physical features of farm production. First, the ownership of farmland is unattractive to capital because it cannot be depreciated, and investment in farmland has very low liquidity as a consequence of the thin farm real estate market. Second, the labour process on very large farms is hard to control because farming operations are spatially extensive. Third, economies of scale are hard to achieve beyond what has already been realized by medium-scale enterprises. Fourth, risks from external natural events like weather, new diseases, and pests are hard to control. Finally, the cycle of reproduction of capital cannot be shortened because it is linked to an annual growth cycle in plants, or a fixed reproductive cycle in large animals. An important exception to this constraint has been in poultry, where there has been considerable success in shortening the reproductive cycle, and this has had important ramifications for the development of capitalist farming as we will see. For all of these reasons we do not expect to see, and have not seen, the wholesale direct takeover of farm ownership by large corporate enterprises employing large, well-controlled labour forces.

ENTIRE CHAIN OF AGRICULTURAL PRODUCTION

The agrifood system, however, is not simply farming. It includes the farm operation, but also the production, transportation, and marketing of the inputs to farming, as well as the transportation, processing, and marketing of the farm outputs. While farming is a physically essential step in the entire chain of agricultural production, the provision of farm inputs and the transformation of farm outputs into consumer commodities have come to dominate the economy of agriculture. Farming itself now accounts for only about 10 percent of the value added in the agrifood system, with 25 percent of the food dollar paying for farm inputs and the remaining 65 percent gained by transportation, processing, and marketing that converts farm products into consumer commodities. At the beginning of the century the value added on the farm was around 40 percent of the total food dollar, and many of the inputs were produced directly on the farm in the form of seed, draught animals, feed for the animals, manure and green manure for fertilizer, and family labour. For the most part these inputs are now purchased in the form of commercial seed, tractors, fuel, refined or synthesized chemical fertilizers, and machinery and manufactured chemical substitutes for labour. Thus, it is the production of farm inputs and the transformation of farm outputs that have provided an opportunity for industrial capital to capture profits in the agricultural sector.

Like any other industrial processes, the production of farm machinery, chemicals, and seeds, and the turning of threshed wheat into a box of breakfast cereal at the supermarket checkout counter are completely controlled by capital and its demands. The problem for capital, however, has been that sitting in the middle of the transformation of petroleum into potato chips is an essential step, farming, in the hands of two million petty producers. They cannot be dispensed with, they own certain essential means of production whose ownership cannot be concentrated (land in particular), and, while they are economically rational, they consume their surplus rather than turning it into capital. Agriculture is unique among all the sectors of capitalist production by possessing at its productive centre an essential process organized around large numbers of independent petty producers. It is as if the spinning of yarn, the weaving of cloth, and the sewing of garments were in the hands of a few large capital enterprises (as they are), but the dyeing and finishing of the raw woven material were unavoidably the exclusive province of hundreds of thousands of home producers who bought the unfinished cloth and sold their product to clothing factories.

9

Green Agriculture

INTRODUCTION

Agriculture and allied sector contributes 24% of the total GDP and provide employment to around 67% Indian population. Use of chemical fertilizers and pesticides have played a positive role in increasing agricultural productivity and in making India self-sufficient in food grain production. Yield of food-grain in India increased from 644 k.g. per hectare in 1966-67 to 1636 k.g. per hectare in 2000-2001(www.indiastat.com) i.e. this registered an impressive increase by around two and half times. This was mainly brought about by a more than 12 fold increase in the consumption of chemical fertilizers (from 1.1 million ton to 13.56 million tons) during the same period. This apart, inorganic chemical use in agriculture has also contributed towards increasing productivity of cash crops.

After this prolonged dependence on inorganic and mineral components for agriculture growth there has been an increasing demand for rethinking agricultural growth strategy. Agriculture sustainability, soil degradation (soil productivity and soil structure), bio-diversity, impact on human health and on environment as a whole are the some of the concerns that are being raised for reviewing part of the agricultural growth potentials based on the current strategy. Search for alternates with a focus on long-term sustainability of agriculture has been enhanced in the last decade. In developed countries the initiatives towards greening agriculture have been prompted both by market attractiveness as well as state support activities.

Usage of bio-fertilizer & bio-pesticides, organic farming, Biodynamic farming (Planning Commission, 2001) low input agriculture, permaculture, sustainable agriculture, integrated farming practices (integrated pest management and integrated nutrient management), are some of the practices that are being espoused by proponents both in developed and developing countries. All these practices have evolved as alternatives to chemical use in agriculture keeping in view the increasing demand for green agriculture products across the world. This growing demand for green agriculture products is both a constraint as well as window of opportunity not only for the agriculturists but also for producers, suppliers and traders of agriculture inputs (fertilizer, pesticide etc.) and outputs.

As a result of increasing domestic and international demand for greening agriculture across the countries both economic and non-economic actors such as institutions/organizations, industrial and trading firms, farming communities, civil society and their representatives played significant role in determining the extent of greening the agricultural production activities. In order to understand the greening initiative in agriculture contextually, it is pertinent to have an overview of trends in chemical inputs into the agriculture in India. Hence in the following we present a brief overview of the same in Indian agriculture.

THE MEANING OF GREEN AGRICULTURE

The word "Green," as applied to gardening, farming, and agriculture, is, in my estimation, pretty poorly understood. In fact, the implications of a "Green" food supply probably cause, or will cause, some uneasiness even to those of us who like to think in terms of "organic" or "sustainable" food production. The word "Green" in the broadest conception of the culture of food production, means a great deal more than just a set of techniques or methodologies applied to food?growing processes. Knowing that you can obtain excellent how?to information from a multitude of sources—how to garden, how to farm, how to spin and weave, how to put together your own alternative energy equipment—I am going to tax your patience by talking about history and culture.

The biggest word we have for all the aspects of food production is not agri-method or agri-technique or even (though becoming so) agri-business but agri-culture. And the Latin roots of agriculture come from words which refer to field cultivation. The words "cultivate," "culture," and "cult" are all related and they reflect, sequentially, the practical methodologies, the social relations, and the spiritual or religious significance of how we grow, share, and are grateful for the Earth products which enable us to live, individually and collectively. A good part of the Green criticism of and even anger toward agribusiness lies in the recognition that agribusiness represents a kind of how-to, chemically-reductionist tunnel vision—concerned only with maximized yields and maximized profits. ("Get big or get out," as former Secretary of Agriculture Earl Butz proclaimed.) Agribusiness does not care about and is not interested in the cultural meaning and spiritual content of non?market food production.

The word "market" is a kind of cultural gate between agriculture and agribusiness. That is, if you believe that the only real way to understand food production is in terms of the market, you are by definition in the camp of agribusiness. Now you may argue for an "organic" agribusiness because of concerns about chemicals or genetic engineering. That would put some jolly

Green make-up on the agribusiness giant. We do realize though, without quite grasping or being able to predict the dimensions, that an "organic" agribusiness would necessitate smaller farms and more labor, for if you can't poison a field or critters into submission, you've got to spend more hands-on time out there cultivating. And that means being more labor-intensive.

This is a small step toward what "Green" means and requires and we haven't yet even come close to Wendell Berry's insight that it is "the subsistence part of the agrarian economy that assures its stability and its survival." Agriculture is relatively new in human history. It originated in the Late Stone Age and—so the anthropologists, archaeologists, and historians tell us—derives primarily from gatherers, not hunters.

That is, women as the foragers for roots, nuts, fruits, and seeds discovered and were the first to practice the intentional planting of crops for human consumption. The abundance so produced, over time, resulted in a number of things. It led to villages which were stable in location, because people no longer had to follow the food, as it were. It led to a greater density of human population. It made hunting somewhat obsolete because of the eventual domestication of various animals, including the cow, donkey, chicken, and horse. This abundance has led one great historian, Lewis Mumford, to talk of "the Golden Age" of the agrarian village prior to the rise of civilization proper.

GREEN REVOLUTION IN AGRICULTURE

Irrigated agriculture in the Indo-Gangetic Plain of northern India and Pakistan is critical to the food security and livelihood of some 500 million people in an area that includes the largest concentration of poor people in the world. The agricultural sector, employing over 70% of the population and accounting for up to one-half of total GDP, is the key to income growth and poverty alleviation in this area. Beginning in the mid-1960s, most of the region has experienced rapid agricultural growth as a result of widespread adoption of Green Revolution technologies, accompanied by investment in irrigation and market infrastructure. The Green Revolution was spearheaded by the introduction of modern varieties of rice and wheat, the two major crops in the Indo-Gangetic Plain.

In recent years, there has been considerable concern regarding productivity growth and sustainability of these irrigated agricultural systems. Use of modern inputs is now high over much of the area, especially in the Indian and Pakistan Punjabs, and further intensification of input use is providing low returns at the margin (Byerlee, 1992). Intensification of cropping, especially in the wheat-rice cropping system which dominates much of the area, has also resulted in apparent degradation of the resource base in the form of salinization, over-exploitation of ground water, deterioration in soil physical and chemical properties, and pest and disease problems. Despite this evidence, there is considerable controversy about the aggregate performance of the agricultural sector and the severity of resource degradation. There does seem to be clear evidence that the growth rates of yields of rice and wheat have slowed sharply over the past decade despite continuing increases in input use.

However, good performance in non-food crops, such as cotton in Pakistan and oilseeds, fruits and vegetables in northwest India, and in livestock, may have offset the slowdown in foodgrains. Crop yields are, however, only a measure of partial factor productivity. Total factor productivity (TFP) growth has been widely used as a measure of the overall performance of the agricultural sector. TFP compares an index of output changes with an index of input changes, so that the residual growth of productivity is attributed to technical progress, changes in the quality of inputs, and changes in the physical and economic environment. Over the longer run, experience from industrialized countries suggests that TFP in the agricultural sector should grow at 1.5-2% per year, with one-third to two-thirds of that due to investment in agricultural research and extension.

Recent estimates of TFP for agriculture in Pakistan and northwestern Indian provide conflicting conclusions. In Pakistan, two studies indicate negative TFP growth in the post-Green Revolution period, especially in the Punjab. By contrast, a study by Khan (1994) concluded that TFP in Pakistan has grown sharply in the period 1980-92, at a rate of 2.1% annually, suggesting that the Pakistan agricultural sector has performed well in recent years. There is little recent evidence on aggregate TFP growth for northwest India although the few studies available generally indicate positive TFP growth.

The above studies vary widely in the coverage of inputs and outputs, methods of valuing inputs, the index procedure used to estimate TFP, and the level of disaggregation. These differences make it difficult to draw meaningful comparisons and reach definite policy conclusions. Moreover, none of the studies have attempted to quantitatively disaggregate productivity growth into the effects of technical change, education and infrastructure, and especially, degradation in the quality of land and water resources.

Better estimates of productivity growth and sustainability under intensification are critically important to food security and poverty alleviation in the Indo-Gangetic plain, the breadbasket of northern India and Pakistan. This paper reports comparable estimates of TFP growth in the Indian and Pakistan Punjabs since the advent of the Green Revolution. Considerable resources were invested to collect data on individual crop and livestock products, inputs, and prices at the district-level in both states.

To avoid the problem of aggregation across heterogeneous regions, TFP is estimated separately for specific agro-ecological zones defined in terms of cropping systems. The delineation of similar cropping systems across states enables direct comparisons of productivity trends and helps determine if productivity slowdown and/or environmental degradation is associated with particular cropping systems and ecologies. In Pakistan, where detailed data on resource quality were collected, productivity growth is econometrically decomposed into the effects of technology, resource degradation, human resources, and infrastructure.

AGRICULTURAL GROWTH AND PERFORMANCE

While the above analysis has provided a general view of the impact of economic reforms, this chaper examines agricultural growth and performance in the states of Bihar, Karnataka, Tamil Nadu and Punjab with their attendant policy implications. The yields for various crops in these states differ greatly. While Tamil Nadu had the highest yield in rice, oil seeds and sugarcane, Punjab enjoyed the highest yields in wheat, coarse cereals, pulses and food grains. Karnataka on the other hand is seen to do well in cotton and Bihar performed quite well in pulses and coarse cereals. Further analysis and findings by Kalirajan and others (2001) show that Punjab had made remarkable achievements on the agricultural front while Bihar had remained stagnant in the last two decades, with Karnataka and Tamil Nadu showing moderate achievement. Clearly, differences in physical endowments, climatic conditions and institutional characteristics are some of the reasons for the varying productivity performance. Thus, having across the board economic reforms is likely to work less effectively than state-specific policy measures that enable each state's agricultural yields to reach their full potential. The comparative advantage of each state's agricultural production should be determined and with inter-state restrictions removed, total agricultural output would see a very significant increase.

For example, Karnataka with less favourable soil and water resources should be given incentives to concentrate on agro-processed products and corporate agriculture in horticulture, floriculture and animal husbandry, or to undertake watershed development to help with dry land agriculture. Many studies have indicated that with watershed areas, productivity growth has been mainly due to seed and fertilizer use. Thus, this state has to be given input subsidies for high yielding seed varieties but at the same time, the farmers need to be educated on the over use of chemical fertilizers.

With Bihar, agricultural performance is problematic on many fronts. First, although demographic pressure has increased and agricultural technology has improved, most of the uncultivated land is concentrated in southern Bihar, where irrigation facilities have not kept pace and the soil is of poor quality. Given the physiography of southern Bihar, wells are also unsuitable and thus the dominant mode of irrigation has been through tanks whose expansion and maintenance has been neglected. Second, the infrastructural facilities of Bihar have been lagging as seen by the infrastructure development. Due to infrastructural bottlenecks, availability of modern goods and services has not increased or their supply remains costly or unreliable.

Third, agriculture in Bihar is dominated by small and marginal farmers and the prevalence of mass poverty is largely related to the backwardness of agriculture. Fourth and importantly, the state agricultural policies in Bihar are in dire need of review. The semi-feudal production condition still exists in rural areas and the ineffective protection of tenancy rights has hindered agricultural growth. The slow pace of land consolidation reflects inadequate financial outlays and a shortage of manpower. Kalirajan and others (2001) note that marketing and extension services in Bihar are also rather weak compared to the other states.

Punjab on the other hand, was one of the few states which enjoyed the success of land reforms and the high priority of investment in rural infrastructure. Also, the irrigation base of the small and medium sized farms was comparable to that of large farms. In addition, the Punjab Agricultural University at Ludhiana contributed to the development of new seed varieties.

However, there are clear signs of a decline in crop yields since the 1990s and this has been associated with the increasing use of fertilizers and excessive water use which have increased the unit cost of production as a result of declining soil quality. Hence, care is needed when providing further input subsidies in fertilizer and water use. Another related fact is the steep increase in wages in Punjab and in the absence of productivity increases, the cost increase has affected the profitability of farmers. With Tamil Nadu, the main crop has been rice as this state is blessed with two monsoons. But from 1992-1997, there has been a steady decline in the areas irrigated by canals and an increase in well-irrigated areas while the use of tanks remains an unreliable source of irrigation. However, major improvements in about 10 rice varieties released in the early 1990s can be expected to improve productivity growth in rice production although pests and diseases as well as imbalance in the use of fertilizers are major constraints. Thus Tamil Nadu could do with subsidies of pesticides and farmers should be educated on the more effective use of fertilizers to obtain high yields. Interestingly, the cropping pattern of late has shown increasing substitution of food crops by commercial crops but there is concern that the benefits will reach farmers only with the development of adequate infrastructure such as roads and markets. However, that Tamil Nadu has a higher index than the all India average of infrastructure.

Challenges of Globalization

It is important to realize that globalization poses many challenges to a developing country like India, which had relied on a state directed and regulated policy regime for more than four decades. In moving to a more open, market-based economy there are many transitional problems that the country has to manage. The Government must play a pro-active role in facilitating the globalization process so that the opportunity sets for the economic agents are widened and the adverse effects of globalization are minimized. The Indian Government must also prepare the necessary information base and develop its capacity to articulate India's concerns and policy trade-offs in the international forums for multilateral trade and environmental negotiations.

In addition, the Government should embark on an extensive programme to educate farmers on the need to meet the standards required in the export markets. In fact, India needs to seek technical assistance in creating the capacity for meeting such standards and to consider watershed developments for environmental considerations. Equally important is the need to disseminate information about possible export markets to farmers, so that market access is achieved at minimum cost. Given the requisite information about markets and profitability, the likelihood of farmers investing in post-harvest and processing technologies and storage and efficient transportation arrangements as well as developing supporting infrastructure is very high.

Although the brave and bold move by India to reduce the tariff rate for agricultural products from 113 per cent in 1990-1991 to 26 per cent in 1997-1998 deserves to be applauded, the question of whether India is ready to compete in world markets remains to be seen. The infant industry argument may still hold for India to shield itself from external competition but one can easily question the length of time that is required to that end. Also, a delay in opening up to foreign trade has the danger that local producers may become too complacent and never be ready for competition.

As India opens up externally, it is also expected to face vulnerability in the wider international price fluctuations and thus Acharya (1998) claims that a minimum price support scheme is important. These prices can also act as a signal to adopt modern inputs and invest in yield-raising infrastructure for increasing production. For instance, keeping basic staple food grains at reasonable prices would induce farmers to switch over to high value crops. However, during the 1990s, Kalirajan and others (2001) shows that procurement prices especially for rice and wheat have been increasing faster than the general price level. Such high prices along with guaranteed purchases by the Food Corporation of India have pushed up market prices. These higher prices are partly responsible for the large buffer stocks with the Food Corporation. If this trend continues, India's comparative advantage will be eroded.

With openness and high price instability, unstable export revenue can also be expected. One way of reducing such risk is for India to diversify her agricultural exports. For example, since 1990, even in commodities such as tea, coffee, cocoa and spices, where India is supposed to have a comparative advantage international prices have been unstable. Besides increasing the type of exports to obtain more export revenue, India should also seriously consider exporting more value added agricultural products through agro-processing such as processed vegetables, fruits, fish and meat products given that export or even local demand for basic agricultural products would decline as incomes rise. The move to higher value added activities within the agricultural sector also spells greater opportunities for industrialization and vice versa as borne by Kalirajan and Shand's (1997) findings of a bi-directional relationship between agriculture and industry for most Indian states. On the other hand, Sivakumar and others (1999) establish empirical evidence of high forward linkages of agriculture due to the presence of agro-industries while Satyasai and Viswanathan (1999) show the significance of the spillover effects to the industrial sector via the intensive use of purchased inputs in the agricultural sector.

The lack or slow pace of internal or domestic liberalization is also seen to hinder the possible gains from external or trade liberalization. For example, although central zoning restrictions have been abolished, state government restrictions on inter-state and even inter-district restrictions on marketing and movement of goods still exist in many cases. This interferes with the benefits from crop specialization and economies of scale arising from comparative advantage. The land market is another example of distortion whereby land ceilings exist preventing the operation of large-sized farms. This has led to the emergence of a large number of small economically unviable land holdings. The easy leasing of land should be permitted with assurance of resumption. Yet another problem lies with the insufficiency of credit to agriculture. From 1995-1996, the Rural Infrastructure Development Fund was set up to allocate funds for the completion of projects and the government has committed itself to strengthening the cooperative credit structure through substantial refinancing and restructuring of the Regional Rural Banks. However, as mentioned earlier, due to varying institutional factors in the Indian states, these domestic reforms can be expected to yield quite different results.

Although India missed the opportunity to open up two decades ago, its attempts to do so now must be regarded as better late than never. Others such as Desai (1999) observe that, "the logic of the global economy as well as India's interests dictate that India become proactive in its liberalization policies. India must liberalize not because it has no choice but because it is the best choice".

His lament that India has adopted a 'victim mentality' when it really needs to adopt a 'winner mentality' has become less of a concern as over time, India has shown commitment to stay on the bandwagon of globalization. Having realized that globalization is a necessary but not a sufficient condition for high growth production, India has undertaken economic reforms, both internal and external. However, it must be ensured that these reforms are synchronized so that the pace of both reforms is set right in order to work hand in hand to promote agricultural productivity growth.

Thus, training the farmers and educating them appropriately to change their mindset and reorienting them to take up new activities or adopt foreign technology is of utmost importance. In this context, it is necessary to involve non-governmental organizations in training and mobilizing the rural poor to face the challenge of liberalization.

Also, with domestic economic reforms, more care needs to be exercised to draw up state-specific liberalization measures to maximize their benefits. Lastly, in the implementation of these reforms for successful globalization, one crucial element, not entirely within control is the need for good governance and stability in the political and economic environment. Political leaders who are the ultimate decision makers in these matters need to examine their own role dispassionately.

It is quite apparent that at this relatively early stage, there is little observable evidence of gains to India's agricultural performance after opening up. However, there could easily be benefits that have not yet surfaced, or are yet to be identified and perhaps too difficult or intangible to measure.

Whatever the case, it is highly likely that it is too soon to assess the full impact of globalization and economic reforms. Furthermore, the process of liberalization has been gradual and remains incomplete. For example, the complete removal of quantitative restrictions after March 2001 will have provided an opportunity for Indian farmers to tap world markets and, if they are successful, results should start to become evident soon. Export promotion via the development of export and trading houses as well as effective liberalizing export promotion zone schemes for agriculture are fairly recent measures and only time will tell as to how effective these measures are. Other possibilities such as agro-industry parks for promoting exports are also in the pipeline.

GREEN IMPACT OF AGRICULTURE

The green impact of agriculture varies based on the wide variety of agricultural practices employed around the world. Climate change and agriculture are interrelated processes, both of which take place on a global scale. Global warming is projected to have significant impacts on conditions affecting agriculture, including temperature, precipitation and glacial run-off. These conditions determine the carrying capacity of the biosphere to produce enough food for the human population and domesticated animals. Rising carbon dioxide levels would also have effects, both detrimental and beneficial, on crop yields. The overall effect of climate change on agriculture will depend on the balance of these effects. Assessment of the effects of global climate changes on agriculture might help to properly anticipate and adapt farming to maximize agricultural production.

At the same time, agriculture has been shown to produce significant effects on climate change, primarily through the production and release of greenhouse gases such as carbon dioxide, methane, and nitrous oxide, but also by altering the Earth's land cover, which can change its ability to absorb or reflect heat and light, thus contributing to radiative forcing. Land use change such as deforestation and desertification, together with use of fossil fuels, are the major anthropogenic sources of carbon dioxide; agriculture itself is the major contributor to increasing methane and nitrous oxide concentrations in earth's atmosphere. We noted several times in the preceding that the era of the new sociology of agriculture was initiated in the aftermath of a distinctly new political milieu of agricultural science and research. Hightower *Hard Tomatoes, Hard Times*, although sensational in tone and misleading on several aspects of the public agricultural research system, attracted a great deal of attention from defenders and critics of the land-grant research system. Most importantly, Hightower's book (along with contemporary ecological criticism of U.S. agricultural technology) opened up several new lines of social science inquiry into the structure and functioning of the public agricultural research system.

Probably in no other state did public agricultural science become as controversial as in California after the publication of Hightower's book. The California Agricultural Experiment Station has long had close relationships with industrial farmers and other segments of agribusiness and had been held directly responsible in the eyes of many for developing new technologies, such as the tomato harvester, which had resulted in the rapid loss of tomato farmers and the displacement of hired farm workers. It was thus not surprising that the initial sociological inquiries into the conduct of public agricultural science were conducted in California.

The first set of studies by Fujimoto and others at the University of California, Davis, were directly inspired by Hightower and were largely focused on how agricultural scientists made research problem choices. In particular, Fujimoto and colleagues devoted attention to how research problem choices were determined by "internal" factors (such as tenure and promotion considerations) and "external" factors (such as the influence of agribusiness interests and grants). They found on the basis of personal interviews with scientists that external factors were quite important, though internal factors also exerted major influence on scientists' research priorities.

The second set of studies focused on California was initiated by Friedland and associates, who examined the processes by which agricultural commodity groups organized to exert influence on individual scientists, departments, and the University of California system as a whole in order to ensure the development and availability of technologies that would enhance their profitability. Friedland and associates gave particular attention to how the responses of University of California research officials and scientists came to emphasize mechanization and mechanization-related technology that enabled growers to displace farm workers or to enhance the control of growers over their hired labor force. Based on the results of several years of research during the 1970s, Friedland and Kappel published a short, but still influential, monograph calling attention to biases in the public agricultural research system and outlining mechanisms by which social scientists and groups adversely affected by particular types of agricultural research could achieve changes in the land-grant system.

Shortly after the California studies were initiated, a research group at the University of Kentucky led by William B. Lacy and Lawrence Busch began a programme of inquiry into the sociology of agricultural science that continues to make major contributions to a sociological understanding of the processes by which new agricultural technologies are generated. The first publication of this group, by Busch, was a theoretical critique of the prevailing conception of scientific knowledge by researchers in the diffusion adoption tradition. Subsequent publications by the Lacy-Busch group have focused on a variety of aspects of the structure and functioning of public agricultural research activities.

Two books published by the Lacy-Busch group have had particular impact on rural sociology and the larger social science community exploring agricultural research. The first, *Science and Agricultural Development,* edited by Busch and containing several papers by his research colleagues, was the first book-length treatment of agricultural research in the tradition of the new sociology of agriculture. The chapters in the Busch anthology most germane to the sociology of agricultural science are the following: Randolph and Sachs compared the establishment of medical and agricultural science, emphasizing how a reductionist orientation and technocratic ideology had been developed in each during the process of institutionalization.

Busch and Lacy summarized the existing literature and their preliminary research on sources of influence in research problem choices in the agricultural sciences. Striving to further the social scientific conceptualization of agricultural science and technology as social products, they explored two rival social science views of the social construction and social structure of science. One, the "internalist" view deriving from Merton work, Sociology of Science, essentially holds that the outcomes of the scientific research process are determined by the "norms of science," held by practitioners and reinforced within scientific institutions, and by the very subject matter itself. The other perspective, the "externalist" view, argues that the outcomes of the research process are largely shaped by the institutions, interests, and availability of resources that originate and are determined outside of scientific disciplines and research institutions.

They suggest that both perspectives have a certain validity: "The internal and external positions are best viewed as distinct strata in the life-worlds of practicing scientists. Thus, for each discipline and institution, science is a continuous process of multilevel negotiation regarding the sources of influence on research". A final paper in the Busch collection of particular relevance to the new sociology of agriculture was Busch and Sachs', which employed a world-systems framework derived from Wallerstein to examine the establishment of the agricultural sciences in the First and Third Worlds since the seventeenth century.

Busch and Lacy are known best for their book, *Science, Agriculture, and the Politics of Research* culture, and the Politics of Research, which summarized an ambitious study of research problem choice through the use of historical methods, personal interviews, and sample survey data. Following several introductory chapters that discuss the establishment of public agricultural research, the organization of the public agricultural research system, the social demography of agricultural scientists, and the structure of communication in the agricultural sciences, they explore three major sources of influences on agricultural research: disciplinary, organizational, and extra-organizational. Busch and Lacy found scientific disciplines and disciplinary organizations to be powerful and growing influences over scientists' research problem choices.

Scientific disciplines affect scientists' research choices in a variety of ways, particularly through controlling peer-reviewed journals and establishing the norms adopted by land-grant universities and the Agricultural Research Service (ARS) in tenure and promotion decisions. Busch and Lacy also found dramatic differences among the agricultural science disciplines according to the degree to which they emphasize basic or applied research, the rate at which they accept papers in scientific journals, their norms for scientific productivity, and the degree to which scientists exhibit consensus on what are "hot" topics in the field.

Organizational context — including ARS vs. land-grant university context and the institutional prestige of the land-grant institution — was found to be a significant influence on research problem choices. Busch and Lacy found that ARS scientists were more highly constrained in their problem choices than land-grant agricultural scientists. They also found that institutional prestige was positively related to the degree to which extra-organizational influences and constraints on research choices were manifest. Among the extra-organizational influences, granting agencies and organizations (those hiring scientists on a short-term consultancy basis) were the most direct and pervasive. Whereas there were very large differences among the disciplines in the extent to which their scientists relied heavily on extramural support, Busch and Lacy emphasize that fiscal constraints on public research appropriations are increasing the pressures on scientists to obtain grants and contracts from government agencies, agribusiness firms, foundations, and so on; accordingly, scientists are becoming less autonomous in their research problem choices.

THE RURAL SOCIOLOGY OF AGRICULTURAL SCIENCE

Until the early 1980s, only small communities of rural sociologists were interested in the sociology of agricultural science. This was to change dramatically, however, in the aftermath of a workshop on biotechnology in the summer of 1982 at the University of Kentucky organized by B. Koppel, Busch, and Lacy. Since that conference there has been growing attention to a variety of facets of biotechnology as key focal points of research in the sociology of agricultural science.

This research on biotechnology now represents a large literature, and the approaches developed therein have begun to substantially influence how rural sociologists are exploring other types of agricultural technology such as information technologies. This new line of research has been salutary in yet another respect; sociologists interested in biotechnology have begun to make major progress in uniting the sociology of agricultural science with sociological research on technological change and social impact assessment of new agricultural technologies.

The wide-ranging and potentially integrative character of research on biotechnology, agricultural research, and agricultural change is amply illustrated by rural sociologists having given major attention to no less than 10 major themes under this rubric. With regard to the first, the politics of the evolution of the public agricultural research system, Buttel and Kenney and Kloppenburg found that the basic science discoveries (especially recombinant DNA) relating to biotechnology, the promise of biotechnology to agribusiness firms for developing new input commodities, and federal government concern with the U.S. competitive position in the "high technologies" have rekindled scrutiny of the land-grant and ARS systems for placing too little emphasis on basic biological research. The result was a changing political climate of public agricultural research that has led to rapid expansion of biotechnology programs in the public agricultural research institutions and to a deemphasis on traditional, applied research.

These changes in the research emphases of public agricultural research institutions have stimulated a second area of research in the sociology of biotechnology: industry-university relationships and the changing division of labor between public and private research. Buttel et al., for example, have found that industrial funding of agricultural research at the land-grant universities has increased significantly and this has generated new concerns that agribusiness firms are acquiring excessive influence on research priorities in public research institutions. Lacy and Busch have identified a large number of likely impacts of closer industry-university relationships in agricultural biotechnology: further decline of public plant breeding (especially varietal releases), greater concentration of scientific talent at a small number of land-grant universities, decreased autonomy of public researchers in establishing their own research agendas, increased secrecy, and greater restrictions on the free flow of scientific information.

A third area of research relating to biotechnology has been on agribusiness reorganization. Kenney, Kloppenburg, and Buttel et al. have noted that the rise of biotechnology research has reinforced or led directly to major structural reorganization in the agribusiness industries. With the passage of the Plant Variety Protection Act of 1970, the ability of private firms to acquire patent-like protection for new crop varieties increased the perceived profit potential of seed companies, leading to the incorporation of most major independent seed companies into multinational corporations as wholly owned subsidiaries or joint-venture partners. Biotechnology reinforced this trend, since seeds are projected to become the major commercial vehicle for plant biotechnology; moreover, the basic sciences underlying biotechnology (molecular and cell biology) are quite similar to those that will ultimately be applied in other areas of agricultural inputs such as agricultural chemicals, innoculants, and animal pharmaceuticals. Thus biotechnology has increased the benefits of uniting chemical and plant breeding product lines within integrated multinational firms that manufacture several agricultural input products. Kenney has given particular attention to the role of small venture-capital or start-up companies in biotechnology and has analyzed the processes by which they have developed linkages with large companies.

Goodman et al. have stressed that biotechnology will have two distinct applications in agriculture — "appropriationism" and "substitutionism" — with different consequences for structural reorganization of the food and agricultural industries. By "appropriationism" Goodman et al. mean the use of biotechnology to develop new inputs, such as new crop varieties and animal breeds, for agricultural production. Appropriationist applications of agricultural biotechnology will tend to result in the integration of agricultural input manufacturing with comparable product lines in other life science industries such as chemicals and pharmaceuticals.

"Substitutionism," on the other hand, refers to the use of biotechnology to substitute for existing products. Substitutionism will therefore have the effect of severing existing production "chains," in which specific raw materials are employed to produce particular final products (*e.g.*, substituting single-cell proteins for vegetable proteins as animal feed supplements). The implications of "substitutionist" applications of biotechnology will also include shifting food and fiber production from the farm to the factory (*e.g.*, substituting industrial tissue-cultured thaumatin for cane sugar). Substitutionist applications will tend to lead to integration of the food and fiber processing and chemical industries.

Fourth, several researchers have examined how biotechnology has affected and will continue to affect plant breeding, including but not limited to public-sector plant breeding. Kloppenburg has conducted an extensive historical analysis of plant breeding and the seed industry in the United States. He has argued that the trends under way as a direct and indirect result of biotechnology promise to increase the penetration of agribusiness interests into land-grant university plant breeding programs and to complete the shift, under way since the late 1930s, to curtail the role of land-grant plant breeders in developing public-domain finished plant varieties. Hansen et al. have reached similar conclusions on the basis of interviews with plant breeders and biotechnology researchers across the public and private sectors. Hansen et al. have noted as well that the increased importance of biotechnology in public-sector plant breeding programs, combined with new avenues for agribusiness influence, will tend to result in the demise of public plant breeding efforts for minor crops.

A fifth area of inquiry in biotechnology, agriculture, and agricultural research has been on the role of intellectual property restrictions and other proprietary considerations in affecting research priorities and the nature of the new technologies that are developed. Kloppenburg, for example, has called attention to private-sector plant breeding firms' longstanding concern about the perceived inadequacy of intellectual property institutions to protect their investments in plant breeding. In the absence of such restrictions for sexually reproduced crops prior to 1970, most seed companies focused primarily on hybrid seeds (largely hybrid corn), which, because of their reproductive instability, afforded a "natural," biological protection against farmers saving their own seeds and circumventing annual purchases of seeds from private companies. Kloppenburg has noted, however, that the elaboration of new legal means of privatizing plant research has increased corporate interest in plant biotechnology. Kloppenburg, Kenney, and Hansen et al. have suggested that the role of intellectual property restrictions, combined with the fact that most agricultural input firms have both seed and chemical product lines, may lead to the development of "seed-chemical packages" that tie the farmer more directly to purchased, capital intensive inputs.

Sixth, Kloppenburg and Buttel have begun to explore the relationships between the public and private sectors in terms of recently developed "theories of the state". They have begun this task by conceptualizing public agricultural research and government commodity programs as reflecting contradictions of state intervention.

They note public agricultural research with state intervention has tended historically to be expanded during times of socioeconomic crisis. Agricultural research, however, can be seen to set the stage for future crises in the agricultural economy by creating long-term increases in agricultural productivity. They discuss the emergence and resolution of these contradictions in the context of the Great Depression and the current era of biotechnology and farm crisis.

A seventh area of research pursued within the rubric of biotechnology has been the implications of this new trajectory of research for genetic resources. Kloppenburg and Hansen et al. have argued that the increased role of intellectual property restrictions combined with the new opportunities for private firms in developing new purchased input commodities will tend to lead to a decline in germplasm flows between the public and private sectors. In the absence of greatly expanded public and quasi-public appropriations for germplasm preservation, these trends are seen to threaten the genetic resource base and perhaps to constrain plant improvement in the future. Kloppenburg and Kleinman have also examined the forces, including the rise of biotechnology, that have led to North-South conflict over control over crop germplasm.

Eighth, as noted earlier, considerable attention has been focused on how new biotechnology-related inputs such as bovine growth hormone, herbicide-tolerant crop varieties, nitrogen-fixing cereal grains, and so on might affect the structure of agriculture in the future. Kenney et al. have emphasized that agricultural biotechnologies are potentially quite diverse and that how these new technologies will affect farm structures depends heavily on the research priorities that are established.

Molnar and Kinnucan, basing their work on existing theories of and evidence on technological change in agriculture, argue that larger farmers will likely be the first to adopt — and thus be the major beneficiaries of — the new biotechnologies. Kloppenburg has developed a preliminary ex ante assessment of how new biotechnologies applied to corn might affect U.S. farm structure over the coming decades. Buttel has focused on the first major biotechnology to be applied in agriculture: bovine growth hormone, which can be produced in large quantities by genetically modified bacteria and which increases milk production per cow by up to 25 percent. He has argued that this biotechnology will likely reinforce longstanding structural trends towards a dualistic farm structure. The Molnar and Kinnucan anthology contains several important papers on ex ante assessment of new agricultural biotechnology inputs.

A ninth area of biotechnology-related research has been in providing preliminary assessments of how biotechnology might affect the nonfarm economy of nonmetropolitan areas. Finally, Buttel et al., Kenney and Buttel, and Koppel have done research into the implications of biotechnology for international agricultural research and agrarian transformation in Third World countries.

The sociology of agricultural research, especially its recent focus on biotechnology, has been conducted within the "new sociology of agriculture" tradition to a greater extent than any other area of inquiry, save perhaps for the state and agriculture. In particular, the sociology of agricultural research, although involving some quantitative analysis, has been at the heart of trends in the new sociology of agriculture that have created legitimacy for exploring issues that are largely accessible only through historical-qualitative research methods.

Especially in biotechnology, the impacts of which on farm structure will generally not be manifest for a decade or more, quantitative research based on primary data is virtually impossible, and the quantitative research that has been done often suffers from spurious precision. Given the importance of understanding the likely impacts of these new technologies, the work that is being done is of high priority. Some rural sociologists, however, will feel less than comfortable with a research tradition in which quantitative methodology and the putative precision it affords are largely absent. Nonetheless, the sociology of agricultural research and biotechnology has been among the most innovative and stimulating areas of recent research in rural sociology and will no doubt have a major impact on rural sociological inquiry in the decades to come.

Agriculture and the Rural Community

Perhaps in no other area of the "new sociology of agriculture" is there so much apparent continuity with the pre-Korean War tradition of rural sociological research as in research on farm structure and rural communities. The early rural sociologists primarily conceived of agriculture as one of the major institutional ensembles affecting rural community life and, especially during the 1930s, devoted much of their research effort relating to agriculture to understanding the implications of agricultural structural changes for rural community life and viability.

The conceptualization of agriculture and rural community interactions and the methodologies for exploring these interactions in pre-Korean War and post-1970 rural sociological research, however, exhibit dramatic differences. Whereas most of the early rural sociologists tended to be less interested in agriculture per se and were principally concerned with understanding rural community institutions, contemporary researchers have tended to be more interested in structural change in agriculture and have deemphasized understanding the totality of the fabric of rural social life.

The research tools and data sources of these two generations of rural sociological researchers are also dramatically different. The early rural sociologists relied on rural social surveys, while the bulk of contemporary research is based on county-level census data. A further difference between the early and contemporary rural sociologists of agriculture in their work relating to rural communities concerns the pivot of the latter studies around Goldschmidt's "As You Sow". This is somewhat ironic in that Goldschmidt's study was anthropological and qualitative, whereas the research conducted since 1970 that has drawn on Goldschmidt's work has largely been quantitative and has involved far less "anthropological" detail on rural communities than did Goldschmidt's.

Rodefeld is generally credited with the "rediscovery" and popularization of Goldschmidt's pioneering work among rural sociologists, though Heffernan (1972) made important contributions to elevating "Goldschmidt-style" research to a prominent position in rural sociology. Nonetheless, after

Rodefeld's refocusing of rural sociological attention on the Goldschmidt thesis, Goldschmidt's book has been virtually an obligatory citation in research articles on this topic. Despite some trenchant criticisms of Goldschmidt's perspective and research methodology, Goldschmidt's general hypotheses remain accepted theoretically and thus continue to shape, if not dominate, research on farm structure and rural communities within the "new sociology of agriculture."

Briefly, Goldschmidt's thesis was that there are differences in the quality of rural community life depending upon the social organization of agriculture and its attendant occupational structure. Goldschmidt observed that the quality of rural life in two carefully matched California communities in the San Joaquin Valley differed dramatically and that these differences were largely accounted for by the organization of agriculture.

He argued that Dinuba, a community surrounded by "family" farms, had a greater number of businesses, a larger retail sales volume, a higher per capita income, and a greater diversity of social, educational, recreational, and cultural institutions than did Arvin, a community surrounded by large, industrial, corporate farms. These differences were principally accounted for by an intervening variable — community occupational composition — that was seen to be strongly influenced by the social organization of agriculture and which directly contributed to the dramatic variations in indicators of the quality of life. Goldschmidt reported that the high ratio of laborers to other occupational groups in Arvin accounted for its low level of community viability relative to Dinuba. From these observations Goldschmidt inferred that the rise of large-scale corporate agriculture then underway in California and elsewhere would have adverse implications for rural communities.

Goldschmidt's study received essentially no further attention until it was referred to in a 1968 hearing on corporation farming before the Senate Select Committee on Small Business. LaRose and Sonka and Heady soon reported further data consistent with the Goldschmidt study. In addition to Rodefeld's and Heffernan's efforts to resurrect Goldschmidt's work in rural sociological circles, there was a major research effort by the Small Farm Viability Project in California, drawing largely on Isao Fujimoto's unpublished research, which presented a larger restudy of the Goldschmidt thesis. The results largely supported Goldschmidt's findings of over three decades earlier. Flora and Conboy also reported results consistent with the thesis.

Since that time there have been numerous studies in the new sociology of agriculture tradition that have explored the Goldschmidt thesis in a number of contexts. Several observations can be made on the conduct of recent research in the Goldschmidt tradition, what we might call "neoGoldschmidt" research. First, it must be stressed that contemporary rural sociological researchers have taken considerable liberties with Goldschmidt's formulations. As noted earlier, Goldschmidt's method was a comparative anthropological case study of two communities and eschewed quantitative analysis of the sort that has predominated in the subsequent rural sociological literature. Also, Goldschmidt's analysis focused on several aspects of the property relations and control over labor, such as absentee ownership and use of ethnic minority laborers, rather than farm size, which has been the most important farm structural indicator in the neo-Goldschmidt tradition. In that regard it is important to stress that the "family" farms of Dinuba were by no means "small" farms; in fact, the family farms of Dinuba were generally larger in gross sales terms than most large farms in the Midwest today.

Second, neo- Goldschmidt research has largely focused on a single direction of causality — the impact of agricultural structure on rural communities — and has largely ignored other types of interrelations between agriculture and rural communities. Third, most studies have been based on Census of Agriculture data and have employed countylevel data as a proxy for community characteristics have utilized survey data to infer community quality of life consequences of changing agricultural structures. Fourth, despite obvious major regional variations in farm structures and in rural communities and their economies, there has, until very recently, been a virtual lack of a comparative perspective on farm and community structures.

Goldschmidt and Harris and Gilbert are partial exceptions to this generalization, since they have done research using states as the unit of analysis, though this procedure compounds the aggregation bias problems. Very recently, nonetheless, there has appeared a series of regionally based studies of the effects of farm structure on rural community viability and quality of life commissioned by the Office of Technology Assessment. These studies are Buttel et al., Skees and Swanson, Flora and Flora, MacCannell, and van Es et al. as published in abridged form in Swanson. These five studies were very briefly summarized in the main report. Together they indicate strongly that there are pronounced regional variations in how changing farm structures affect rural communities.

The impacts of farm structural changes on rural communities are generally most pronounced in the Great Plains and industrial agriculture states, less so in the South and Midwest, and very little in the Northeast. There is evidence that the degree to which farm structure affects rural communities is directly associated with the dependence of the rural community on agriculture.

The general model pursued in these neo- Goldschmidt studies has been to estimate coefficients for a causal chain in which farm structural change indicators are seen to affect the composition of the farm population (*e.g.*, as measured by indicators of the size of the farm population, the proportion of full-time hired workers) and the size and composition of the rural population. These clusters of variables are then seen to affect indicators of the quality of life in rural-agricultural communities (*e.g.*, proportion of families in poverty, median family income, per capital retail sales, unemployment rate). This general model is sometimes supplemented by technological factors as antecedents of farm structure, since technological change is considered to be one of the major factors influencing farm structural change. More recently, studies in this genre have gone beyond the cross-sectional census data designs that predominated from the mid-1970s to the early 1980s and have included measures of change, where appropriate, in variables in the model. One limit to studies that use census and other data based on single-year assessments of farm income is that net farm incomes are highly variable over the short-run; for example, as many as 20 percent of farms with annual sales of over $250,000 had poverty-level incomes in 1986.

Heffernan has effectively summarized the general configuration of results in the neo-Goldschmidt literature by noting that "all relevant research to date suggests that a corporate type of agriculture results in a reduction in the quality of community life for at least some people, especially the hired workers in rural communities." As such, the results continue to lend support to the Goldschmidt thesis. But there have recently been several expressions of dissatisfaction with Goldschmidt's perspective and with the generalizations that have been made from research in this tradition.

A major theoretical concern is that Goldschmidt and many researchers working in this tradition have taken large-scale and corporate agriculture to be basically synonymous when, in fact, not all large-scale farms are corporate or industrial-type farms. Much more important than industrial-type farms in farm numbers and shares of sales, assets, and profits are larger-than-family farms. Empirically, as Goldschmidt-style research has been done in the Midwest and other traditional family farming areas, there have been admonitions that large-scale agriculture, especially of a nonindustrial character, does not necessarily lead to adverse impacts on rural communities, though some studies note that increased scale of production could have detrimental impacts on community services.

SOWING THE SEEDS OF THE GREEN REVOLUTION

Realising the disastrous consequences of the growing gap between the rates of growth of population and food production, a vigorous 'grow more food' campaign was launched. The key goal of economic planning shifted to attaining self-sufficiency. The First Five-Year Plan, following food imports in 1951, gave maximum importance to the growth of agriculture. Among the measures initiated in the '50s to stimulate foodgrain production were land reforms, irrigation, fertiliser production, strengthening of research and the organisation of a national farm extension service. Consequently, increased harvests in the mid-1950s relieved the economy. It was around this time that the government adopted the US Land Grant model of agricultural research. Agricultural universities, beginning with the Gobind Ballabh Pant University of Agriculture and Technology in Pantnagar in Uttar Pradesh, were gradually set up in all the states. The Land Grant model entailed linking research with extension, and taking research from the laboratories to the farms through an extension network linked with the university. In 1961, the government launched the Intensive Agriculture District Programme (IADP) to enhance productivity in the irrigated areas. Although the aim was to introduce improved seed along with a package of agronomic practices to take advantage of the available irrigation, the programme did not come up to expectations for want of high-yielding varieties of wheat.

The tall Indian wheat varieties did not provide an economic response to the application of chemical fertilisers and irrigation water, which agricultural scientists were desperately looking for. These had a tendency to fall down or lodge when fertilisers were applied, as a result of which the yields stagnated at less than one tonne per hectare. Even though breeding for improved non-lodging wheat varieties was initiated in the '50s, it did not meet with much success. The missing link was provided by Dr Norman Borlaug, who was then working with CIMMYT in Mexico. A few dwarf wheat strains of spring wheat that he sent to the Indian Agricultural Research Institute in New Delhi, looked impressive. In the next two to three years, a wide range of dwarf material was tested under the All India Wheat Research Project. In August 1964, the then Professor and Head of Department of Plant Breeding and Genetics at the IARI, Dr M S Swaminathan, who later emerged as the architect of India's Green Revolution, proposed the launching of a National Demonstration Programme in farmers' fields. This would verify the results obtained in research plots and introduce farmers to new opportunities opened up by semi-dwarf varieties for considerably improving the productivity.

Says Swaminathan, "When small farmers, who with the help of scientists organised the National Demonstration Programme, harvested over 5 tonnes of wheat per hectare, its impact on the minds of other farmers was electric. The clamour for seeds began and the area under high-yielding varieties rose from 4 hectares in 1963-64 to over 4 million hectares in 1971-72. A small government programme became a mass movement." Subsequently, the government imported 18,000 tonnes of semi-dwarf wheat from Mexico in 1966 to be distributed in time for immediate sowing by farmers in the northwestern regions.

The seeds of the Green Revolution were, therefore, truly sown in the mid-'60s. What followed next is already part of contemporary history. Wheat production rose to 17 million tonnes in 1968, an increase of 5 million tonnes over the highest of 12 million tonnes harvested in 1964. Indira Gandhi officially recorded the impressive strides in agriculture by releasing a special stamp entitled 'Wheat Revolution' in July 1968. The success of wheat was later replicated in rice. Productivity increases were also recorded in cotton, sugarcane, millets and oilseeds.

MEASURES TO AVOID FAMINE

The famine-avoidance strategy that India launched had all the necessary ingredients for a sustained effort. In 1965, the Agricultural Prices Commission was constituted, with the basic objective of assuring fair prices for farm produce. The Commission, an autonomous body, works out the cost of cultivation for 22 agricultural commodities, and suggests a price to the government that includes profit that should serve as an incentive to grow more. The administered price of wheat, for instance, was raised by 71 per cent between 1972 and 1981. Similarly, the price of rice was jacked up by 81 per cent during the same period. Later, to sustain the tempo of the rice revolution, the relative price structure was moved in its favour. In the early phase of the Green Revolution, the price policy favoured wheat, but after the mid-'70s it tilted in favour of rice. For oilseeds and pulses, the reverse was put into action. Relatively high support prices were announced in the early-'90s to make their cultivation profitable.

Perhaps the most significant aspect of the price support mechanism has been the insulation of farmers against a decline in prices. Wheat production was very high in 1971-72, as a consequence of which prices should have declined by at least 25 per cent. Since the government purchased all the wheat that was offered, farmers were actually paid 7 per cent more. Again, in 1982-83, when wheat production was 14.3 per cent higher than the previous year, the prices should have actually declined by 32.5 per cent. With the government purchasing all the wheat that flowed into the markets, however, the farmers realised 18 per cent more.

The biggest achievement of the Indian food policy, dependent more and more on buffer stocking and operational stock holding, has been the avoidance of famine-like conditions. It was with the basic objective of curbing consumption and ensuring an equitable distribution of available food supplies, especially in the deficit areas and among the poorer strata of society, that the Public Distribution System (PDS) was introduced. PDS supplies were also used for the 'food for work' programme as well as other anti-poverty programmes. At present, PDS covers more than 80 million families and the total foodgrains distributed through a network of 40,000 fair price shops accounts for over one-third of the total trade in foodgrain.

AIMING AT SELF-SUFFICIENCY

The Green Revolution was certainly not a stroke of luck. The resilience was the result of progressive farm policies, determination, sustained efforts of scientists and the vigour of the Indian farmers who readily adopted the improved technology. This is what the Rockefeller Foundation had to say about the magnificent strides taken: "... The speed with which India's farmers and scientists suddenly gave their country an approach to an abundant food supply, has never been duplicated on an equal scale anywhere in the world, including the scientifically sophisticated United States." Since the early-'70s, the country has avoided famine even during adverse weather conditions, through a carefully designed food security system involving the maintenance of both substantial grain reserves and an extensive PDS. The devastating drought of 1987, rated as the most severe of this century, affected nearly 155 million hectares. In spite of this, no starvation and hunger deaths were reported. This was essentially the result of excellent food management based on foodgrain reserves built largely from homegrown wheat and rice.

From the late-'60s, the rate of growth in food production generally exceeded the rate of growth of population. Despite the continuous growth in population, the per capita availability of foodgrain increased from 157 kg per year in 1955 to 177 kg per year in 1995, an increase of 15.7 per cent. Food production has risen roughly four times since the days of Independence, from 50 million tonnes in 1950-51 to 198 million tonnes in 1996-97. At the same time, the population has risen from 300 million to 960 million. Milk production too has risen to 68 million tonnes, the highest in the world. Food self-sufficiency was achieved in the late-'70s with the maintenance of a sizeable food buffer. Lately, India has begun to export foodgrain. About 5 million tonnes of rice and wheat were exported during the last few years. Skimmed milk powder amounting to about 5,000 tonnes has also been exported.

The increase in foodgrains was, however, disproportionate. The highest rate of growth was achieved in Punjab and Haryana, where, at a compound growth rate of 4.63 per cent, foodgrain production jumped from 7.23 million tonnes in 1964-65 to reach an all-time high of 30.33 million tonnes in 1995-96. Uttar Pradesh recorded a compound growth rate of 3.01 per cent. Production in Maharashtra, Madhya Pradesh, Bihar, Orissa and the northeastern states continues to stagger.

Tamil Nadu and Andhra Pradesh, on the other hand, have recorded significant increases in rice yields, with production rising to 9.17 million tonnes and 11.58 million tonnes, respectively.

The Green Revolution was aimed at the better-endowed regions. For millions of farmers languishing in the drylands, constituting more than 70 per cent of the cultivable lands, it continues to be a futile struggle. Despite emphasis on dryland farming during the past several decades, the scenario still remains grim.

The undulating topography and the irregular rainfall pattern have combined to aggravate the situation. That the drylands produce about 42 per cent of the country's food shows that the future of farming lies in these areas. Nearly 83 per cent of sorghum, 81 per cent of pulses and 90 per cent of oilseeds grown in the country come from these areas. The poor yields and the fluctuations in production are indications of the scant attention drylands have received from policymakers and planners. The problem of increasing productivity on drylands has serious socio-economic implications. With every passing year, the gap between the farmer's yields in irrigated areas and in the dry farming region is widening. One year of drought is enough to push a farmer into a deep well of poverty for another two to three years. Drought is a recurring phenomenon in arid and semi-arid areas. Fifty years after Independence, life for millions of people somehow surviving in the drylands continues to be worse than before.

DECLINE IN PRODUCTION

Thirty years after the dawn of the Green Revolution, Indian farmers are realising that their love affair with intensive agriculture is on the decline. Despite a bountiful monsoon for the tenth year in a row, harvests are not as plentiful as expected. However, with no let-up in the ever-increasing rise in population, Indian agriculture is once again at the crossroads.

Alarm bells have been ringing for quite some time. For nearly a decade, agricultural production had stagnated. The spectacular yield growth recorded in the post-Green Revolution years in Punjab and Haryana has receded into history. Of the multiple problems confronting agriculture, rapid fragmentation of land holdings is keeping pace with increasing population. In 1976-77, the average size of the holdings was estimated at 2 hectares, while in 1980-81, it came down to 1.8 hectares. Today, it stands at a mere 0.2 hectares. The total number of land holdings in 1981 were around 89 million; today these have crossed 100 million. By the turn of the century, the average land holding will come down to 0.11 hectares. It is quite obvious that with such small land holdings, Indian agriculture cannot adopt high-tech farm practices.

Basking in the afterglow of the Green Revolution, farming and agriculture have ceased to attract serious attention. In 1995-96, foodgrain production slumped to minus 3.60 per cent. In 1997-98, it further declined to minus 3.70 per cent, the worst-ever since the heady days of the Green Revolution. Foodgrain production in the frontline agricultural states of Punjab, Haryana and western Uttar Pradesh, comprising the country's food bowl, has decelerated.

The miracle that began with wheat was replicated in rice. This is where the cost-benefit ratio has turned negative. The downward spiral in food production continues through the southern regions of the country. Tamil Nadu, another Green Revolution area, is under tremendous strain from intensive cultivation.

In Karnataka, the negative trend in yield levels of all food crops, barring cotton and sugarcane, are all too apparent. Farming in Karnataka can be clearly separated in two distinct classes, the 'creamy layer' of corporate agriculture occupying the fertile and irrigated areas, and the remaining low productive tracts at the mercy of the rains, constituting the tiny and small land holdings. Kerala too, picturesque and verdant with tropical forests, is passing through its worst-ever crisis on the foodgrain front.

As if this is not enough, the number of landless in rural areas too has been multiplying over the past few decades, at an estimated rate of two million every year. The rapid expansion in the number of land holdings, with less than a hectare under plough, has meant an additional 6,00,000 farmers every five years or so. The negative terms of trade for agriculture and the declining public sector investments in farming are indices of the sluggishness in this sector. The Economic Survey for 1996-97 has further sounded a warning. The compound growth rate in foodgrain production at 1.7 per cent between 1991 and 1996, is lower than the annual population growth of 1.9 per cent. The survey adds that even a marginal fall of three to four per cent in foodgrain output can cause the prices of essential commodities to escalate sharply, necessitating government intervention by way of resorting to imports. This happened in 1993 and again in 1996.

The philosophy of agricultural planning is changing. Gone are the days when the nation's emphasis was solely on attaining self-sufficiency in foodgrain production. Gone also are the days when farmers were newly independent India's heroes, revered for their role in keeping hunger and starvation at bay. Today, at a time when food production struggles to keep pace with the burgeoning population growth, farmers are being asked to diversify, produce crops that are suitable for export and to compete in the international market.

With the promise of cheap food available off the shelf in the global market, the focus has shifted from agriculture to industry, trade and commerce, from the small and marginal farmers to the agri-processing companies, which alone can bring in investments and add value to produce. The restructuring of the Indian economy too has put the emphasis on agri-exports. Cultivation of staple food is being replaced by cash crops, tomatoes in place of wheat, durum wheat (for bakery purposes) replacing wheat as a staple diet in Punjab and Haryana, flowers in place of rice, and so on. In the coastal areas, private enterprise is taking away the fish catch, depriving the local communities of a livelihood and their only source of nutrition. In Kerala, vast tracts of forests and paddy fields have been converted into rubber, coffee and coconut plantations. Every year, about 25,000 hectares of good paddy land are being diverted for non-paddy purposes. The structural transformation is not only peculiar to Kerala. It is happening in almost all the states. Commercial crops are eating into the fertile land tracts meant for growing essential foodgrains, thereby making the states rely more heavily on the PDS. The diversion of good agricultural land, which in any case is limited, to commercial farming and even industries, is further restricting the ability to grow enough foodgrain.

IN THE ERA OF WTO

Kerala has already been at the receiving end. Over the past few years, ever since economic liberalisation became the development mantra, the state has been flooded with cheap and highly subsidised agricultural imports, throwing its agrarian economy out of gear. Whether it is the import of palm oil or rubber or coffee, almost every aspect of the state's socio-economy has been negatively impacted. Coconut prices have crashed, down from Rs 10 to Rs 2. Rubber has plummeted from Rs 60 to Rs 16 and coffee from Rs 58 in 1999 to Rs 30 per kg in 2000. Even spices have not been spared, with pepper prices falling from Rs. 2,600 to Rs. 1,300 per quintal in the consecutive period. Kerala is not alone. The fallout from the emerging global trade paradigm is being felt all over the country, though not in the same magnitude. Before we go any further however, it is important to understand how the market rules play, and then try to understand the implications for Indian agriculture.

It is now official. Six years after the World Trade Organisation (WTO) came into existence on January 1, 1995, the anticipated gains for India from the trade liberalisation process in agriculture are practically zero. The ministry of agriculture as well as the ministry of commerce have admitted that the hopes from an international regime that talked of establishing a fair and market-oriented agricultural trading system have been belied.

It is time to draw a balance sheet of the gains and losses that accrued from the implementation of the WTO's Agreement on Agriculture, which had incorporated three broad areas of commitments from member states, namely in market access, domestic support and export subsidies. The underlying objective of this was to correct and prevent restrictions and distortions in world agricultural markets. Six years later, it has been established that these measures have only protected the farmers and the farming systems of the developed countries. On the other hand, the trading regime has ensured that developing countries take time-bound initiatives to open up their domestic markets for cheap and highly subsidised imports of agricultural commodities.

Index